ECOSYSTEM SUSTAINABILITY AND HEALTH
A Practical Approach

Improving the health of people and animals, and improving the health, integrity or sustainability of ecosystems are laudable and important objectives. Can we do both? There are no ecosystems untouched by human activity, and there are worrying signs that the world's ecosystems are reaching the limits of their ability to adapt to human impacts. Drawing on fields as diverse as epidemiology and participatory action research, philosophy and environmental sciences, ecology and systems sciences this book is about searching for solutions to complex problems to produce a new science for sustainability.

DAVID WALTNER-TOEWS is a veterinary epidemiologist in the Department of Population Medicine at the University of Guelph, Canada, specialising in the epidemiology of zoonoses, diseases of animals that can be transmitted to humans who live with them, share their environments, or eat them. He is also founding president of the Network for Ecosystem Sustainability and Health (NESH), which brings together some of the most cutting edge thinking in complexity and sustainability with local community development in many parts of the world.

ECOSYSTEM SUSTAINABILITY AND HEALTH

A Practical Approach

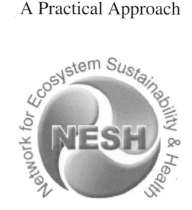

DAVID WALTNER-TOEWS

University of Guelph

CAMBRIDGE
UNIVERSITY PRESS

PUBLISHED BY THE PRESS SYNDICATE OF THE UNIVERSITY OF CAMBRIDGE
The Pitt Building, Trumpington Street, Cambridge, United Kingdom

CAMBRIDGE UNIVERSITY PRESS
The Edinburgh Building, Cambridge, CB2 2RU, UK
40 West 20th Street, New York, NY 10011–4211, USA
477 Williamstown Road, Port Melbourne, VIC 3207, Australia
Ruiz de Alarcón 13, 28014 Madrid, Spain
Dock House, The Waterfront, Cape Town 8001, South Africa
http://www.cambridge.org

First published 2004

Printed in the United Kingdom at the University Press, Cambridge

Typeface Times 11/14 pt.   *System* LATEX 2$_\varepsilon$ [TB]

A catalogue record for this book is available from the British Library

Library of Congress Cataloguing in Publication data
Waltner-Toews, David, 1948–
Ecosystem sustainability and health : a practical approach / David Waltner-Toews
p.   cm.
Includes bibliographical references and index.
ISBN 0 521 82478 8 – ISBN 0 521 53185 3 (pb)
1. Ecosystem management.   2. Environmental health.   I. Title.
QH75.W365   2004
333.95′16 – dc22   2003068837

ISBN 0 521 82478 8 hardback
ISBN 0 521 53185 3 paperback

This book is dedicated to two people who have challenged me and prodded me into unexplored realms of theory and action I could not have imagined: N. Ole Nielsen, the Green Dean; and James Kay, the Dirk Gently of bio-complexity.

Contents

Figures

Note: All figures © NESH 2003, by permission, except where otherwise stated.

Tables

Introduction

Improving the health of people and animals as well as improving the health, integrity and sustainability of ecosystems are both laudable and important activities. Can we do both? Clearly, if we wish to have health in the future, then the integrity of ecosystems, which make our lives possible, is relevant. To say we can have sustainable population health without sustainable ecosystems is like saying that we can have a sustainable, healthy heart without a sustainable body, which gives it life and meaning. Yet linking health and ecosystems grammatically – a common and generally well-received notion these days – will do little to link them in real life. Some people would argue that the only ecosystem with integrity is one with no people in it. These people seldom use the word health because they think that health involves value judgements, and integrity is value-free. If anything, integrity is more value-laden, and indeed legally moralistic (which is why it attracts some environmental regulators), than health. Nature may well be value-free, but there is no way to evaluate our status in nature, or to talk about progress, without reference to values. It seems best to some of us to accept this and try to deal with it head-on. There are, quite frankly, no ecosystems that do not, in one way or another, bear the imprint of human meddling.

Conversely, it is possible to achieve population health, at least in the short run of a few hundred years, by radically restructuring and perhaps endangering ecosystems. People of European descent have done this for decades – draining swamps, chlorinating and diverting waterways, cutting down dark and dangerous forests and replacing them with carefully tended crops or regimented tree plantations. We now have improving indicators of human health world-wide, largely as a result of this strategy. We also have worrisome signs that the world's ecosystems may be at the limits of their ability to adapt to this radical restructuring. As I consider the losses of our fellow species on this planet, I wonder if the improvements in health come at the expense of an impoverishment of well-being.

Between these two extremes, a body of theory and practice has developed. While acknowledging the tensions between the health and well-being of the various species with whom we share the planet and the ecosystems which nurture and give us life, this new field of inquiry also seeks to find the interactive, relational space that is our common future. Converging from disciplinary bases as diverse as epidemiology and participatory action research, philosophy and environmental sciences, ecology and systems sciences, a new, integrative, place-based science for sustainability, or post-normal science as Funtowicz and Ravtez have dubbed it, has emerged. In this fertile and hopeful ground, a new kind of practice is taking shape.

Much has been written in the scholarly literature about the intellectual basis for this new science. These theoretical developments provide the basis for a gen-eralization of sustainable action. However, the specifics of what form that action might take, especially for health practitioners, have yet to be brought together in a coherent way. Just as medical diagnostic techniques are not the same as those used for health promotion, the methodologies used to *understand* the ecosystems in which we live may be inappropriate for *promoting* ecosystem health. Without the right tools, many practitioners simply fall back on the same old toolbox. We have a thermometer, so the problem must be temperature. Where's the rectum? We have a net, so the problem must be fish. Where is the river? We have data on income. Therefore the problem must be economic. Where is the mathematical model? The point is not that measuring temperatures, fish populations, or incomes is unneces-sary but that they only acquire meaning in context – and people will only act upon information which they think is meaningful. Furthermore, if we want to promote realistic national and international policies, then we need to have some idea of what might actually get us where we want to go. Moral umbrage can only get us so far – and often in the wrong direction.

This book is about searching for solutions to complex problems. The health, agricultural and ecological problems we face in the year 2004 are qualitatively different to the problems for which standard scientific, medical and political tools and programs were designed. Given the messy nature of the dilemmas and con-tradictions facing us, there can be no single recipe, and no definitive set of tools. However, some approaches, ways of thinking and ways of doing seem to be more useful than others. The ecosystem approach, as defined and used by researchers and managers of the International Joint Commission within the Great Lakes Basin, is one such approach (Allen *et al.* 1993; Kay *et al.* 1999). Grounded theoretically in complex systems, and practically in participatory research and adaptive man-agement, the ecosystem approach is a way of working with people in such a way that measurements are given meaning by understanding their context, or rather, that both measurements and action emerge from the context.

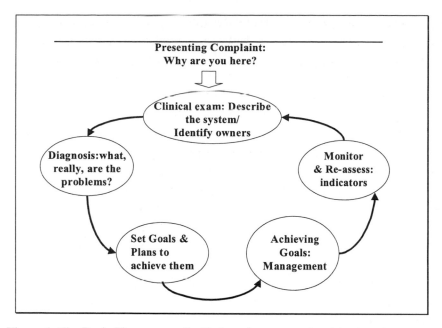

Figure 1 The Basic Figure: a medically based assessment and treatment process.

This book is designed primarily for use by practitioners, that is, those who wish to understand and improve – in a sustainable fashion – human, animal and ecosystem health. This certainly includes health practitioners – veterinarians, physicians, nurses and public health workers. It will also be of use for those in fields such as agriculture, environmental management, wildlife biology, city planning, food safety and international development, who are working alongside health workers in addressing these complex problems.

Just as there are many ways to describe the complex reality in which we live, there are many ways to describe the process of assessing and improving the systems of which we are a part, and this text will present a selection of those. None of them by itself captures the whole complexity either of the system or of the process. Because this text is addressed to health practitioners (in a broad sense), the health management process, as outlined in Figure 1, which I shall refer to throughout the book as the Basic Figure, serves as a useful starting point. Although it will lead us in some significantly new directions, and will in fact undermine itself to the point that we must conclude that medical approaches are not only inadequate but counterproductive, this process begins in what hopefully is familiar territory for health practitioners. It draws on the medical diagnostic process and the herd health management model used by veterinarians when assessing the health of groups of farm animals.

This process was originally designed for examining and treating individual people or animals. Someone comes into a doctor's office with a presenting complaint – a headache, perhaps, or a fever. Or a farmer calls the vet because her cow is not eating. Or a father visits a community health nurse because the baby won't stop crying. Once the patient or client is through the door and has had a chance to express his or her complaint, there follows a clinical examination of the patient, a diagnosis of what the problem is, some suggestions as to what might be done about it, and then some actions and follow-up. At the farm-animal herd level, the principles are basically the same, except that the presenting complaints may have to do with low reproductive rates or poor growth rates, and economic (making enough money) and social (having time to spend the money) considerations begin to get mixed into the goal-setting. Once we begin talking about ecosystems and communities, the problems get even more complicated – indeed they get (in technical terms) complex. That is, they resist understanding by any single method or set of methods.

These complications of the basic diagnostic and treatment process alter, in some fundamental ways, our understanding of disease and health, and the skills required to prevent one and promote the other. In a standard diagnostic or herd health management process, there would be little discussion about who the owners are, and we are usually taught that the nature of the problem is somewhat independent of who defines it. A reproductive or disease problem – so many people believe – is a problem no matter what the owner thinks. This is actually false, which has led to many of the battles between, say, European farmers and American drug companies about what constitutes improvement in animal health. When we are tackling ecosystem health, the nature of the problems and the ownership of the system interact even more closely than they do at herd level; in fact, ecosystems have many owners, not all of them human, and what some of the owners see as problems, others see as solutions. This is why, in the Basic Figure, the description of the system (which comes from the clinical exam) and the identification of owners appear in the same circle. Ultimately, we will also move to an understanding that there are close relationships among the diagnosis, the description, and ownership. These considerations lead us beyond basic disciplinary science to a kind of public, integrated, contextual science in which the various actors and owners are part of the process of generating knowledge and critically evaluating it. In many ways, this has much more in common with Paulo Freire's 'problem-posing' within a total context than it does with conventional academic and business science. The main differences between Freire's action-education and the ecosystem approach is that the latter is explicitly rooted in complex systems and ecological perspectives, as set out by Kay *et al.* (1999).

Despite all the complications, however, the underlying processes of assessment, goal-setting, action and reassessment (monitoring) hold true. If you get lost in the forest of complications in the chapters that follow, it might be useful, periodically, to refer back to the Basic Figure. Once we have worked through the entire process, I will (in Chapter 6) present a revised version, the Adaptive Methodology for Ecosystem Sustainability and Health (AMESH), which incorporates our new understanding of complex systems, eco-social change, and how human communities can live sustainably and convivially on the planet. James Kay, Tamsyn Murray, Cynthia Neudoerffer, myself and several others have developed AMESH and tested it in ecosystem health-type projects around the world. It is our hope that AMESH will become the new starting point, the new baseline, for investigating and resolving the complex problems presented to us by communities and the ecosystems of which they are integral members.

An earlier version of this book was titled 'Ultimate Patients'. For reasons of clarity in marketing, the title was dropped, but I think it is worth reminding ourselves that the eco-social systems, of which we are a part, are the 'ultimate patients' whose pathologies we seek to limit, and whose health we seek to promote. We also need, in the midst of the urgent agendas besieging us, to find ways to 'think like ecosystems', to develop a kind of 'ultimate patience'. One approach to the catastrophic ecological changes occurring around the world is to panic, rushing into Draconian, undemocratic measures. These will surely backfire. Another is to take a more measured, deliberate, directed approach, perhaps like a veterinarian or physician in an emergency. There are important things we need to do but the level of uncertainty and the stakes are so high that rushing is unlikely to improve the situation, and may well make it worse. In ecosystem health, as in animal and human health, our first aim is to do no harm. I hope this book can contribute to achieving that goal.

1

Presenting complaint

Within any health profession, we begin to examine a person, animal, community or ecosystem when we have some inkling that something might be amiss. Usually, someone comes to the practitioner with a complaint: the animal has diarrhoea, the person is having trouble breathing, the water smells funny, there are dead ducks along the shoreline. This is called the 'presenting complaint'. Certain symptoms and signs characterize this complaint. Symptoms are what a person or animal feels (headaches, depression); signs are what can be measured (temperature, heart rate, dead bodies). We tend to think that a dysfunctional ecosystem might have signs but no symptoms; however, ecosystem ill-health may be manifest by symptoms in the people and animals living there. For instance, poet Leonard Cohen captured the feeling of dis-ease between external events and internal feelings in one of his songs when he said it 'looks like freedom but it feels like death'. In general, presenting complaints have to do with symptoms, and practitioner responses have to do with signs. This book will tend to focus on signs, but the process we finally arrive at in Chapter 6 is designed to improve symptoms as well.

What are the clinical signs?

While those who are primarily concerned with environmental management might struggle with the need to find a coherent framework within which to define, evaluate and promote 'progress', we might ask why health practitioners need to be bothered with this. Don't we already have a successful global medical and health enterprise, suffering perhaps from under-funding, but, where money is available, bringing longer and healthier lives to everyone? Do we have any evidence that something might be wrong?

The answer to this question is more complex than it appears at first glance. While disease management and mortality prevention have been very successful in the latter half of the twentieth century, we are beginning to see signs that this success, and success in other fields of human endeavour such as agriculture, is actually creating serious new problems. The signs we are seeing may be early warnings that we are pushing the world's ecosystems to the limits of their capacity to absorb human impacts. A. J. McMichael discusses these in his book, *Planetary Overload: Global Environmental Change and the Health of the Human Species* (1993). Even if this is not the case – even in the most Pollyanna, the-world-is-okay scenario – the signs we are seeing indicate problems that are serious in their own right, and worth addressing.

Initially, we start with a list of the kinds of signs that, we believe, reflect systemic problems. Some of these are clearly at a particular scale (it's hard to see a hole in the ozone layer as a local problem), while others could be at any or all scales (species loss, for instance). Here is a starter list:

hole in the ozone layer
soil erosion
resistance of insect vectors to pesticides
loss of non-target insect species
loss of non-target birds and mammals as a result of attacking disease vectors
frogs dying
chemical spills
trees dying
acidification of lake water
dead, disappearing or deformed fish, dolphins, seals . . .
people getting sick or dying
contamination of drinking well or tap water
increases in the size, number and nature of foodborne diseases
West Nile virus outbreaks
Hantavirus outbreaks
floods/ droughts/ sudden rainfalls/ more storms or hurricanes (sudden weather events).
antibiotic resistance in microbial populations
irrigated soils become too salty to use
epidemics of malaria, obesity, starvation . . .
vultures dying in India
botulism epidemics in Merganser ducks on Lake Erie

The list, of course, is almost endless. Given such a list – which in any given context will be finite and limited – how can we begin to work our way from these signs back to the shape and size of the patients we are dealing with?

The first step is to organize the signs into some sort of coherent framework. Some signs pertain to particular spatial scales of system. Water-related problems (contamination, scarcity), for instance, may indicate ecosystem stress or dysfunction at a watershed level. Epidemics of disease are characteristics of populations. Floods and droughts might be related to regional climate changes.

Some signs actually reflect the context for others. Thus, water contamination with pathogens might be a sign of an ecosystem problem in its own right, but may also be seen as a contextual risk factor for human diarrhoea. Disease epidemics may occur because particular wetlands dry up, which may reflect global climate change. So, we might focus on one scale, but quickly find ourselves moving between scales for causal variables (larger scale) or explanations of process (smaller scale).

Another way to organize the signs is by system. Thus we might look at the water system (hydrological cycles), food system (food webs, agrifood system organization), nitrogen cycle, and so on. We might also consider various kinds of pathology: dysfunctions characterized by broken feedback loops (farmers producing for markets without regard for their natural resource base), or unresolved conflicts between an invasive species and the long-term inhabitants (people versus old growth forest, for instance). This way of organizing signs requires greater knowledge of a situation than we might have when we start. Thus, classification by pathology is often retrospective (or, if it is too early, it becomes a sort of pathological classification, creating problems by the way it structures the situation; declaring water contamination to be a water system problem opens some doors to possible solutions but closes others, such as agro-ecosystem management). On the other hand, we know a lot about many of the problems we are dealing with, and an *a priori* classification can help us look for patterns. Only be aware that the classification is a human construct – useful but dangerous.

For some situations we already have sufficient understanding to group clinical signs into broad diagnostic categories. At this point in the eco-health process, these must be seen as tentative diagnoses to guide further in-depth investigations in the pursuit of something more definitive. At least five such systemic diagnostic categories can be created for framing our thinking about emerging infectious diseases.

1. Disease treatments don't work

Many disease-control programs are no longer effective. In fact, one could argue that disease treatments are causing disease. Microorganism and parasite populations are rapidly developing resistance to a wide array of antibiotics and pesticides. Both the range of drugs to which these organisms are resistant, and the proportion of organisms that are resistant, are increasing. This rising tide, globally, of multi-resistant organisms and pesticide-resistant insect vectors is the direct, unintended, result of therapies we use to control or eliminate them. One short-term response to these 'counter-attacks' is simply more of the same – more vaccines, more drugs, more pesticides. In some ways, this is like responding to successful guerilla warfare by proposing bigger conventional armies and weapons. I suggest that it is time to ponder the wisdom of our bio-military metaphors and linear causal thinking, to address the flaws in reasoning and tactics we have employed to date, and to use our much vaunted intelligence as a species to find more creative solutions.

2. Health promotion causes disease

Success in programs which manifestly promote health in some dimensions – such as improvements in agriculture to address food shortages – have had unintended negative effects on other aspects of health, such as disease. Talking about creating 'supportive environments for health' is simplistic. It would be possible to create a large mall that is supportive to health (filtered air, lots of food, exercise gymnasiums, music). In a sense, industrialized countries have created a healthy 'mall' by externalizing costs to the poor and vulnerable. Some water management programs have had devastating effects by favouring several tropical diseases. Dams are built to generate electrical power, to control flooding, and to generate wealth (all of which are demonstrably supportive of health). Nevertheless, they also expand or create new habitats for flora and fauna which cause disease, and remove sources of natural renewal from farmland (Hunter *et al.*, 1982). In Bangladesh, epidemic Kala-azar (lcishmaniasis) has occurred in populations living within flood control embankments (Minkin *et al.*, 1996), and malaria epidemics, 'mad cow disease' and cyclosporiasis have all been associated with aggressive agricultural programs (Waltner-Toews, 1999). Improving the outdoor environment by providing trails and parks, and encouraging people to use them, has resulted in improved physical and mental health in those members of the population who can avail themselves of these amenities. However, these same activities are associated with an increase in a range of diseases such as Lyme disease and West Nile virus infection.

If increasing populations of ducks by creating artificial wetlands can be seen as improving population health (in a Darwinian sense it's at least increasing survival), then the millions of ducks that die each year of botulism in those artificial wetlands can be seen as victims of a disease caused by a health program. On one of our field trips as part of the veterinary Ecosystem Health Elective, we studied one wetland where more ducks died than were born – a nursery turned, in a kind of Stephen King twist of plot, into a mortality sink.

3. Disease control causes disease

Same scale

How can disease control cause disease? This is most obvious in food-borne diseases, where industrialization and centralization, which quite naturally accompanied regulations on canning and pasteurization to control botulism and brucellosis, have been associated with large-scale epidemics of diseases like salmonellosis. This is because the consolidated system has larger ecological niches for bacteria (more cows in one place, more volume of milk mixing) and longer transport distances. Imposition of food safety programs developed in industrialized countries with good, expensive, energy intensive infrastructure, on poor southern countries with bad roads and unreliable power sources will likely worsen the situation considerably rather than improve it. A study of small-holder dairying in Kenya by Amos Omore of the International Livestock Research Institute, for instance, suggested that the best way to ensure a safe supply of milk was to encourage the widespread practice of boiling milk, and support hygiene programs for small producers, rather than promoting centralized pasteurization plants. In North America, policies and practices which encourage a voluminous and cheap supply of food serve, on the one hand, as a preventive against starvation. On the other hand, they also undercut the economic and ecological sustainability of farmers, and are associated with a whole new array of nutritional and disease problems associated with obesity.

Cross scale

Current health and disease control programs often work against each other across organizational scales. Problems are solved at an individual level but become major problems at a regional or global level. Thus, saving children through vaccination without concomitant programs in education, nutrition, agriculture and sustainable livelihoods undermines the health of whole communities and condemns them to slow and painful death and disintegration (McMichael, 1993). Indeed, the tension between sustainable population health, which requires a certain death and replacement rate, and individual health, for which death is the ultimate negative outcome, has no solution within current biomedical models (Waltner-Toews, 2000a). The idea that death and maybe even disease might in some sense be important for sustainable health cannot even be conceptualized in a normal biomedical framework.

At a more mundane level, we have the absurdity of governments in some industrialized countries giving away groundwater to private companies, who then wrap it in plastic, sell it back to the original owners of the water (the citizens of the country) under the pretence that this is good for their individual health. Even if the water in the bottle could be demonstrated to be superior to tap water, it would still have major negative consequences for population health because of the energy and resources required for manufacture and disposal of the bottles.

Drawing inferences about populations based on studies of individuals is termed the atomistic fallacy, and is widespread and widely tolerated in epidemiological studies. Ironically, the converse fallacy – drawing inferences about individuals from population studies – is vigorously guarded against. What this means is that all efforts are focused on finding individual determinants of disease, and the broad systemic conditions – the very conditions which determine whether or not healthy human communities are sustainable – are largely, by design, ignored. We are obsessed with eating behaviours leading to obesity and heart disease, but largely ignore the obesigenic eco-social systems that nurture and encourage those behaviours.

4. Disease control causes ill-health

Disease control programs may not result in outright disease, but may undermine health in more subtle or convoluted ways. They can disrupt ecological systems that make health possible. Thus we are faced with the dilemma that DDT is useful in bringing malaria under control, but at the same time endangers the integrity of the interactions among insect pollinators, birds and food production which make sustainable livelihoods and health possible. Secondly, the consolidation of the agrifood system which created niches for pathogenic microbes was also one of the essential forces which changed the way rural communities were organized, and created considerable ill health in social and community terms – at least in the transition. Thirdly, and less obviously, food supplementation, vaccination and drug treatment programs based on a biomedical model can undermine the ability of people to adapt resourcefully to their own environments. They do this by reinforcing the notion that it is appropriate for outside experts to determine which outcomes – among many possible competing ones – are appropriate, which responses are 'correct', and who should carry them out. Physicians and veterinarians, who are well equipped to diagnose and treat, are in general very poorly trained to promote health, which requires negotiation and adaptation.

5. Biomedical disciplinarity causes blindness and inhibits effective sustainable health promotion

Current disciplinary-bound approaches to health, which focus on biomedical and personal behavioural issues, inhibit health researchers and workers from addressing the real causes – which reflect irreducible interactions among economics, politics and ecosystems.

A 1992 report by the Institute of Medicine (IOM) in the United States reflects the general consensus on the reasons for disease emergence. The authors of the report identified half a dozen forces which were resulting in the emergence of new diseases and the resurgence of old ones (Table 1.1; Lederberg *et al.*, 1992). Rudolf Virchow, in a report to the Prussian government 150 years earlier regarding a typhus epidemic in Upper Silesia, identified causes that are similar to those of the IOM (Table 1.2). The authors of both identified the major causes as being social, environmental and political.

Table 1.1 *Factors in emergence of
new diseases*

1. Human demographics and behaviour
2. Technology and industry
3. Economic development and land use
4. International travel and commerce
5. Microbial adaptation and change
6. Breakdown of public health measures

Source: Lederberg *et al.* (1992).

Table 1.2 *Some of Virchow's recommendations to the Prussian
government regarding the typhus epidemic in Upper Silesia, 1848*

1. Political reform and local self-government.
2. Education
3. Economic reform
4. Agricultural reforms, including development of cooperatives
5. Road building
6. Requirement that professionals such as teachers and physicians speak the
   language of the local people.

Source: Adapted from Drotman (1998).

However, while Virchow's recommendations are overtly social and political –
and hence based on the evidence – those of the IOM do not reflect the evidence
presented at all (Table 1.3). They are at best technical, and at worst seem self-serving
and unrelated to the evidence. Nowhere in the recommendations do the IOM authors
discuss altering the economic and political causes of disease emergence. Based on
the evidence presented, one would think that health practitioners should be making
strong health representations to organizations like the World Trade Organization
and the World Bank, not on how to clean up the disease mess after the fact, but
on how to prevent the mess in the first place. This is one of many instances where
we can see that the ideological lenses through which health and disease are studied
constrain the opportunities to find solutions. This, if nothing else, should raise a
warning flag that those who study disease are not necessarily well-equipped to
promote health, and that new modes of thought which can incorporate multiple
perspectives are required.

Finally, in ecosystem health work, it is often most informative, useful and effective
if we classify problems by stakeholder group: whose problems are these? Who
cares? Who wants to have something done? It is important to involve members

Table 1.3 *Recommendations for action by the Institute of Medicine's Committee on Emerging Microbial Threats to Health*

1. Development and implementation of more effective state, federal and global surveillance systems.
2. Expansion of National Institutes of Health (NIH)-supported research on agent biology, pathogenesis and evolution, vectors and their control, vaccines and anti-microbial drugs.
3. Generation of stockpiles of selected vaccines.
4. Expedite pesticide registration for vector control and stockpile those pesticides.
5. That NIH give increased priority to research on personal and community health practices relevant to disease transmission and education 'to enhance the health-promoting behavior of diverse target groups'.

Source: Adapted from Lederberg *et al.* (1992).

of the affected communities and stakeholders in this classification process, often (although not always) by holding workshops. This is so that the 'problems' and issues can be framed within the cultural world-view of the people in the system. By doing this, we situate the problems in such a way that not only desirable, but also feasible, resolutions can be found. Since we don't actually know who the patients or problem owners are at this point, we need to include all interested parties. Such workshops use standard participatory action research techniques (see Pretty *et al.*, 1995) to list and organize the issues. This leads us directly into the next stage.

What do these clinical signs mean?

If the first step in any examination is to document signs and symptoms, then the next step takes us on the journey of interpreting them. Our interpretation of clinical signs is based on our understanding of how reality is structured – in medical terms, our understanding of physiology and organ systems. As Dustin Hoffman's enigmatic character in the film *The Messenger* explains to Joan of Arc, what she saw was a sword in the grass. How she interpreted that sword – a sign from God rather than merely evidence that a soldier had thrown away or lost his sword – was entirely dependent on her understanding of reality.

The ecosystem approach used in this text is based on an understanding of reality in terms of complex systems theory. Many alternative views exist, ranging from traditional cultural myths to experimentally based scientific theories; complex systems theories have provided a way to accommodate many of these without negating them. In this way of looking at the world, the messy eco-social reality in which we live – indeed which encompasses the entire biosphere – can be described in terms of SOHOs[1] – Self-Organizing Holonocratic Open systems. Such systems can be described according to a variety of criteria. For example, they are in a state far from thermodynamic equilibrium, kept alive by a constant influx of energy, and cannot be understood using any single set of standard mathematical or scientific models. In the course of their normal development over time, they may go through sudden transitions between states. They do not necessarily have one preferred stable state. In this section, I will discuss only three SOHO characteristics which are relevant to our interpretation of the clinical signs above, and which are helpful in determining prognosis and possible resolutions. For those interested in further details on such systems, the book by Casti (1994), the paper by Kay *et al.* (1999), and James Kay's website (www.jameskay.ca) provide accessible doorways into the literature.

The three characteristics that I will discuss are: feedback loops and their consequences; holonocracy and its consequences; and multiple perspectives and their consequences.

Feedback loops, self-organization, attractors and surprise

The interactions in SOHO systems can be represented as a mixture of positive and negative feedback loops. Many of the pathologies I talked about in the previous sections are related to these loops. For instance, people engage in various economic activities – such as clearing land for agriculture, irrigation, mining, house-building – in order to make money to improve the quality of their lives. Wealth generated by these activities may be used to build better roads, schools and sewage disposal facilities. People who have more schooling may be better able to solve social and public health problems – at which time they may see that some of the activities which made the schools possible (cutting down trees and draining swamps) may themselves be identified as problems, undercutting sustainability and restricting future options. Agricultural activities or manufacturing may, for instance, result in greater pollution of the water supply and the environment, heavier stress on energy use, and general deterioration of the ecosystem. Some diseases may be prevented when swamps are drained or dams are built, even as habitats for new ones are created.

In natural SOHO systems, it appears that, as high quality, useful energy (referred to in the thermodynamics literature as 'exergy' to differentiate it from the more general term 'energy') and information are pumped into the system, the feedback

---

[1] Arthur Koestler (1978), probably in a tongue-in-cheek reference to a 'socially active' area of London, used the acronym SOHO to refer to Self-regulating Open Hierarchic Order; the designation I am using is a modification by Henry Regier of one proposed by Kay *et al.* (see www.jameskay.ca), and reflects more recent complex systems terminology.

loops become organized in such a way as to make more effective use of the entering resources, build more structure, and enhance their own survivability. It is this combination of feedbacks, boundaries and openness which results in what is called self-organization. Self-organization is necessary for life to occur. All living things – organisms, ecological systems, eco-social systems – must remain both bounded, with a set of internally relatively stable interactions, and open to receiving resources and energy, and dumping waste, if they are to remain alive.

Some elements in any ecosystem are more tightly connected than others, and more essential to their mutual well-being and/or the well-being of the system overall. The importance of connections is not determined by sector or perspective. In other words, this is not simply a matter of setting social or economic priorities on agriculture, health, business, social and environmental issues. The dynamics of the complex interactions in the socio-biosphere itself, the flows of useful energy, resources and information related to patterns of self-organization may mean that relationships with, say, food may be crucial and irreducibly necessary for everything else, even if they are seen to be trivial economically or socially. Thus, activities that enable a community to make more effective and elaborate use of natural resources, exergy and information are likely to have a greater impact on the viability of a given population than health care activities.

As already suggested, these feedback loops in SOHO systems tend to organize themselves in certain patterns that are coherent and relatively resistant to change. 'attractors' These patterns are sometimes referred to as 'attractors' (there are some contradictory definitions of attractors in the literature; my intent here is not to argue about the language, but to focus on the phenomenon). Most ecosystems – because of the energy and resources available to them – seem to have a propensity to fall into a certain limited set of possibilities. Despite advertising claims to the contrary, not everything is possible, and we cannot all become whatever we want. We – and the ecosystems and societies we live in – are comprised of physical elements, which constrain our possibilities. Lakes may become benthic or pelagic, depending on temperature, nutrient loads and water flows. Agrifood systems may shift from localized, small-holder systems (similar to benthic systems in that they are driven by local, close-to-the ground forces) to massive corporate systems, driven primarily by top-down market forces. The possible states open to any system comprises its canon.

One of the key lessons we learn by viewing reality in terms of attractors is that notions, for instance, of human or animal rights, are only meaningful in context, and that the context is created by feedback loops which occur from acts of responsibility to a collective whole. This responsibility is called citizenship. Relationship is everything. Furthermore, any notion of attractors – and hence any notion of citizenship – which is not defined in terms of eco-social wholes, with information, energy and materials flowing among the elements, is also meaningless. Eating, for instance, is not just a social or an economic act; it is first and foremost an ecological act, whereby people ingest certain parts of their (or others') environments. But moving from place to place (transportation), defecation, sewage disposal and burial rituals are all also ecological acts and profound acts of citizenship responsibility, since they contribute to the self-organizing patterns within which human rights can be defined.

Certain actions in this eco-social system of which we are a part may trigger transitions to new attractors. In the transition from one attractor to another, relationships and species that were peripheral to the old system may suddenly become central or take on new roles. If invasive species have altered the composition of available diversity, then the system may not flip back to exactly what it was before. That is, in an age of rapid global movement of plants, microbes and animals, some changes may be irreversible. Human intentionality and creativity can push or alter the constraints, which may result in new system states – or just in general disintegration. In the latter case, if life is to continue, some new sets of mutually supportive interactions need to arise.

Changes between system states may be quite sudden. These 'flips' between attractor states have been well described for both social and ecological systems, but not well described for the complex eco-social systems that must form the basis for sustainable health. Nevertheless, ideas of thresholds and breakpoints are well known in both the epidemiological and ecological literature. Disease organisms increase to critical levels at which time the probability of adequate contact increases to the point where the epidemic explodes. Diseases such as HIV-AIDs may occur at high endemic levels or low endemic levels, depending on the systemic interactions they are part of, with little gradation between. Similarly, ecosystems can exist in different steady states, reaching critical points and then suddenly reorganizing. During these reorganizations there may be drastic changes in species composition and diseases. Research on ice cores from Greenland has indicated that global temperature changes of the order of 5–16 °C have occurred over mere decades during global climatic changes in the past. Such 'flips' are attributed to the crossing of thresholds of temperature required to keep global ocean currents moving in particular ways. These kinds of threshold effects, which are followed by catastrophic changes between attractors, have been demonstrated for a variety of systems, social as well as ecological (Casti, 1994; Kay *et al.*, 1999).

This means that gradualist views of disease changes in relation to climate, for instance, may be very poor grounds on which to base organizational response plans, and hence will undermine the ability to adapt and respond to stress. It also means that standard epidemiological notions of causality are not only problematical but perhaps nonsensical for those outcomes embedded in complex systems. The occurrence of the disease is embedded in a complex, stable set of interactions (an attractor) in which changing one 'cause' will likely have no effect, or may even have unintended negative effects. Indeed, as I will emphasize later, many of the causes that are the most important determinants of our current state are actually in the future! Farmers plant crops and modify landscapes in anticipation of weather and markets. We build houses and transportation systems in anticipation of future demographic patterns. The expected future thus dramatically changes the present landscape we live in, which changes the possible futures available to us.

While these dynamics and the cascade effects of key variables, species or actions may provide opportunities to facilitate major changes, we are not sure what these influential 'hot buttons' are, nor can we be sure that we would like all the changes that occurred if we hit the button. For instance, if we are concerned about urban respiratory diseases, we may want to start by putting speed bumps on all city streets, and narrowing them to discourage car traffic, while at the same time investing in cleaner public transport. This would likely result within a few years in cleaner air, less respiratory disease, and healthier people who walk more. Of course this might also result in the loss of income from car-related activities, higher unemployment (which has its own negative health effects), and a change in the physical structure of the cities and in the structure of the national economy. Paying farmers for managing landscapes sustainably as well as for producing commodities would change the entire structure of rural communities, migration to cities, international trade, patterns of food-borne disease and global economic power. Not all of the changes would be seen universally to be good. While the exact outcomes associated with radical systemic changes in such complex systems cannot be predicted, an informed public could at least see the general shape of the system options.

Holonocracy and contradictions

Historically, most health practitioners have been trained to think primarily in terms of individuals: how to examine a sick person or cow, for instance. While this is very useful as a starting point, however, it is rarely useful for coming up with solutions to the problems detected.

Cows are members of a herd, and the characteristics of the herd – such as what it is fed, levels of immunity and available housing – will strongly influence the state of health of the cows in it. Herds are also populations in a local ecosystem, which can influence what kinds of pasture and water sources are available. They are, as well, members of larger 'communities of interest', which nowadays are referred to as industries (as in the cattle industry, the swine industry), which often determine whether animals are vaccinated or not, shipped around the country, crowded inside or left to run loose, all of which will influence the health of the herd and the individuals in it.

Dogs, cats and people are functionally members of human families; families together comprise both physical neighbourhoods and various kinds of communities of interest. Individual people may become sick because of the housing conditions determined by their families, as well as the behavioural and physical relationships among family members, including non-human animals. Families may have problems because air in the neighbourhood is polluted, there are no jobs available, and violence is tolerated and encouraged by advertising, religion or television.

At each layer, in any one of these hierarchies, one can look outward to the levels of a nested hierarchy above, and inward, to the sub-systems within. Arthur Koestler (1978) spoke of reality as being Janus-faced, like the two-faced Roman god. An individual person, then, is both a whole in herself, with individual characteristics that define her, and a part of something larger, both socially and ecologically. This is true of a farm, a family, a copse of trees, and a flock of ducks as well. Koestler referred to each one of these as a 'holon', and the nested hierarchy of which they are a part, as a 'holarchy'.

Some writers simply use the term 'nested hierarchy' in lieu of 'holarchy'. I prefer the language derived from Koestler for two reasons. First, the term 'holon' has meaning only in relation to a holarchy. If we revert to 'nested hierarchy', we have no obvious word to refer to the whole parts (people, family, etc.) of which the nested system is comprised. Secondly, the language of hierarchy carries considerable baggage, most of it inappropriate for the situations we are considering. The way we govern ourselves, for instance, tends to be hierarchical – with national governments above state and provincial governments, with local village governments at the bottom. At the very top are multinational private and public corporations and military-industrial complexes. The lower levels of government are not part of the higher levels of government. They are something else. If national governments disappeared, the local governments would still be there. If a herd of cows or a city community disappears, then, by definition, the cows and the people also disappear.

Recently, environmental scientist Henry Regier has coined the term 'holonocracy' to more accurately reflect the mutual power relationships across scales. I like the term better than Koestler's original, since it provides an ecologically grounded counterpoint to terms such as democracy and technocracy, which seem to have impoverished views of power and nature. For these reasons, I shall be using the terms holon and holonocracy in this book. (We should also note that most political hierarchies are hierarchies of the more traditional type, and that by using the general term holonocracy, I may be guilty of mis-classification as often as if I used the word hierarchy. Such are the limitations of language to describe reality! Please differentiate in your head as you read.)

It should already be apparent that holonocracies are not necessarily objectively verifiable, independent things. They represent ways of looking at a complex world. Thus, a farm or a family can be seen in an ecological, an economic, a social, and various other kinds of holonocracies. Yet these holonocracies are related to each other by virtue of the fact that they represent ways of thinking and talking about the same complex reality. Koestler spoke – metaphorically I assume – of the tendency of holonocracies to arborize (grow up and branch like trees across many layers) and to reticulate (create networks at any given level). Both these characteristics are important for understanding some of the signs of pathology we see in the world today.

Allen and Hoekstra (1992), in applying hierarchy theory to ecology, distinguish between scale-defined levels of observation (small areas versus larger areas) and criteria for observation, which may be applied at various spatial or temporal scales. These criteria – organism, population, community, landscape, ecosystem, biome and biosphere – are what ecologists use to determine which relationships they wish to focus on in any given ecological observation. The criteria are something like a holonocratic perspective on reality in that they focus on functional relationships, even though criteria such as the biosphere, landscape and organism tend, for pragmatic reasons, to be associated with particular scales of observation. I say 'something like' because the analogy is not exact. Indeed, Allen and Hoekstra argue that an exact, all-inclusive definition of hierarchies is less important than the acknowledgement that scale and perspective, context and content are relevant to understanding nature, which is more complex than any of our models. It is important to keep in mind that we are not talking about describing with mathematical precision simple systems like computers, cows or human bodies. Rather, we are trying to devise ways of thinking about the complexity we live in that will yield insights into how we might achieve sustainable health.

Multiple perspectives

Complex feedback loops have both positive and negative effects, and the same action may have very different impacts at different levels in a holonocracy. Thus different people will look at any situation – and evaluate it – differently. Where one person sees the excitement of economic activity, another person sees deforestation; where one person sees disease control by draining swamps, another person sees loss of wildlife and the loss of clean water which the wetlands once provided through natural filtration; where one person sees disease control through metal roofing, another person sees increased economic and environmental costs and less comfortable houses. Furthermore, migration of people out of flood zones may be seen as a catastrophe, or a response to catastrophe, at the local level, but simply adaptation at the larger regional scale. This means that, as the scientific description gets better, the problems are not necessarily resolved, only clarified. We will come back to this repeatedly in this text, since this notion – that there can be several, equally legitimate, understandings of reality – complicates immensely our efforts to promote sustainable health.

Diagnosing disease, negotiating health?

By looking at the clinical signs of global eco-disease in the context of SOHOs, we can begin to make sense of them. We can begin to see why, through ignoring feedback loops and attractors, or holonocratic organization, undesirable, unexpected outcomes may result. It would be like ignoring the family of origin if you are a psychotherapist, or pretending the disease condition of a herd has no impact on the cattle in it, or ignoring the cardiac side-effects of a drug used to promote weight loss. We are used to thinking that way at organism and even herd levels. We now need to begin thinking that way when we talk about sustainable public health, agriculture and ecosystems in general. What complicates matters when we are talking about complex eco-social systems is that there is no single textbook description of health. While there are clearly biophysical constraints and opportunities – there are ways to kill the planet – there is more than one way to 'live right' in the biosphere. Hence health is a negotiated construct within biophysical constraints. Exactly how we do that is one of the greatest challenges facing us as a species. I discuss this more fully in the chapter on setting goals for management (Chapter 4).

Who is the patient? Using clinical signs to define the boundaries

The clinical signs, which are leading us to a deeper understanding of the underlying dynamics and structures of ecosystems, are also helping us to define the patient. The shape and size of the holonocratic ecosystems we are working with are determined by the shape and size of the mess we are trying to resolve.

It may seem odd to speak of defining a patient based on the clinical signs. This is because most people – not just disease care providers – tend to think of health firstly in terms of organisms. The organism (person, animal) is brought to our attention, and the clinical signs are interpreted in terms of what we know about that type of organism and how we think it should 'normally' behave. For ecosystem health problems, however, we are faced first with the clinical signs: animals are dying, people are getting sick, towns are flooding, hillsides are eroding, maple sugar trees are dying. What the boundaries of these problems are, or how they might interact with things that we don't consider to be problems (growing food, building homes, creating employment), is not immediately apparent. This has prompted systems scientist Peter Checkland to talk about a problematic situation, rather than a problem per se. Russel Ackhoff, another well-known systems scientist, has called this, simply, a 'mess'. What we are faced with is a situation in which various issues interact, some of which we (or some other people) are not happy with.

In exploring the clinical signs we can begin to see the shape and size of the patient. For instance, suppose we are faced with an outbreak of *Salmonella* DT104, which is resistant to multiple antibiotics, on a particular farm. The simple bio-medical response is to find a different drug, which will be effective in curing the patient(s), and/or to find other ways to contain the disease. What if we think about this more ecosystemically, however? For one thing, we recognize that antibacterial resistance has become a global problem. Furthermore, some researchers at the United States' Centers for Disease Control have suggested that these virulent and resistant bacteria emerged as the result of antibiotic use in aquaculture in Asia, connected to the movement of infected breeder stock or contaminated feed ingredients; other researchers suggest that human travellers may have carried the bacteria to different parts of the world. In both cases, the 'patient' to be considered in resolving the problem is no longer just a particular farm, but a structure of animal rearing, trade and antibiotic use embedded in a multi-layered, globalized, agrifood system (Angula and Griffin, 2000; Waltner-Toews, 1999, 2001). Similarly, some researchers recently proposed models that explained the emergence of *Salmonella enteritidis* as a major human pathogen (carried by poultry but not causing any serious problems in them) as the result of the vacating of a particular ecological niche through the removal of *Salmonella gallinarum*, which was attacked by veterinarians because it was a pathogen of poultry (Rabsch *et al.*, 2000). In this case, the patient may be more restricted, bounded, perhaps, by the nature of the poultry industry and treatable, probably, only within those bounds.

Certainly, a fish die-off in a stream may be the result of a toxic spill, and a die-off of trees may occur because soil around their roots has been washed away, both suggesting local solutions. However, in general, we should be thinking in terms of feedback loops, holonocracies and how they arborize and reticulate, and then focus back on what might appropriately be done locally – and what requires regional or global regulation. The fish may be dying from bacteria that are able to proliferate because certain streams are getting warmer due to climate change or multiple sources of urban effluent. The local trees may be dying from acid rain, which is the result of industrial development and regulatory policies.

In late 2000 and early in 2001, an unprecedented drop in vulture populations was reported in India. The causes of this 'virtual extinction' event were unclear,[2]

---

[2] By 2004 many researchers had concluded that the vultures died from ingestion of carcasses which had residues of diclofenac, a painkiller widely used as a palliative in cattle and buffaloes in India.

```
┌─────────────────────────────────────────────────────────────────┐
│                                                                 │
│        Defining the patient: What scale? What Perspective?      │
│                                                                 │
│                                     Perspectives                │
│                            Biophysical   Economic  Social/ Health│
│                            Gender  Ethnicity  Age  Social Class  │
│           Hierarchies &                                         │
│           Holarchies                                            │
│                          ───────────────────────────────────   │
│           Field/ Individual                                     │
│           Farm/ Family                                          │
│           Sub-watershed/ Community                              │
│           Watershed/ Municipality                               │
│           Bio-region/ State or Province                         │
│           Ecozone/ Nation – International                       │
│           .                                                     │
│           .                                                     │
│           .                                                     │
│           Biosphere/ United Nations                             │
│                                                                 │
└─────────────────────────────────────────────────────────────────┘
```

Figure 1.1 A matrix for classifying the boundaries of an ecosystem 'patient'.

but the systemic consequences included increases in populations of feral dogs, possible increases in rabies, anthrax and other zoonoses as dead bodies were left unscavenged. Even religious impacts were pondered, since the Parsees of Mumbai rely on vultures to scavenge human bodies in their Towers of Silence. While the presenting complaints were often local, the size of the 'patient' expanded outwards to include a whole eco-social system of the Indian sub-continent, which had evolved with vultures as an essential component.

The clinical signs, then, that we are faced with, help us to define the boundaries, and the holonocratic structure, of the system we will need to examine. Thus, in ecosystem health, unlike in other fields of health, we begin with a disease, then move to define the nature of the system we are dealing with, and then we determine the owners. This is quite the reverse of how a normal medical examination works. I will throw an extra complication in here, which I shall return to later: the 'boundaries' to which I have been referring are determined by the sets of relationships in which the presenting problems are embedded. This is only one of many ways to talk about boundaries and, for issues of sustainability and health, may not be the best way. If sustainability is to build on local contextual co-evolutionary history – and if terms like 'invasive species' are to have any meaning, which they surely do – then the most important boundaries to pay attention to in this debate are spatial. (For a discussion of many of the issues relating to boundaries, see Midgely, 2000, 2003.)

One of the first things we can do as we examine the clinical signs and the perceived problems that they reflect, is to classify them by scale (individual, community, region, etc.), and perspective (socio-economic and cultural, ecological). Figure 1.1

shows a matrix which demonstrates some of the elements involved. Scale, in this case, refers to the spatial units with which we are primarily concerned. These units (holons) have their own distinctive internal feedback loops and rules of operation: households or bio-regions, for instance. Perspective refers to the range of relevant views on the situation (economic, by class or gender, biological, etc.) that we must consider.

As we define this patient by scale, a question arises: what is the appropriate spatial scale at which to address ecosystem health issues? If we look at general principles, we see that what we want is a scale that transcends the individual, so that the eco-social feedback loops become visible, and because global survival is not about survival of individuals, but of the context which makes survival of individuals possible. On the other hand, if we go too 'high', we become removed from the real needs of local ecosystems and people, and it becomes impossible to motivate *(absolutely!)* meaningful action. Ultimately, ecosystem integrity and health are defined within local contexts, usually around the size of geographic communities and watersheds. This will vary from place to place (think of grasslands and oceans versus mountain valleys, for instance), but the general principle holds. Izak and Swift (1994) propose that we need to investigate at least one level above, and one level below, the level of interest. Thus, if we want sustainable, healthy communities, we would also consider households and regional ecosystems. In Kenya, we found this useful for investigating the health of rural communities (Gitau *et al.*, 2000); however, in other areas somewhat different approaches are needed. In general, we want to focus our activities where they will do the most good in terms of patient health, which means looking at the entire holonocracy and then zeroing in.

I have spoken so far as if social and ecological holonocracies are commensurate. This can be misleading, since social, political and ecological boundaries rarely coincide. This must be acknowledged and dealt with head on when undertaking this work. Thomas Gitau, for instance, created a table to display social and ecological holonocracies (Table 1.4) (McDermott *et al.*, 2002). As well, he identified levels in the social holonocracy where formal governance and organizational structures existed, and – at the village level for instance – where they didn't. This is important for setting up sampling schemes and is also essential, later, when we identify actors in the system and design strategies to bring them together.

Lack of formal governance structures may be an opportunity for change. In Kenya, the villagers were asked to set up agro-ecosystem health committees to manage project activities. There was only one constraint on these committees: they had to comprise equal numbers of men and women. We discovered later that this changed a whole range of men–women interactions in the village.

From the picture that emerges from our holonocratic diagrams, we can begin to identify the actors and owners in this mess. We can also begin to look at how the

Table 1.4 *Nested and other hierarchies in Kenya*

Biophysical	System boundaries	Examples/types	Policy makers/ managers	Human activity perspective
Geo-climatic	Geographic and climatic features	Arid, semi-arid, highland, coastal, basin	Government of Kenya	Nation
			Provincial administration	Province
			District administration	District
Agro-ecozone	Geology, climate, vegetation, agriculture	Forest zone Tea-dairy zone Coffee-tea zone Marginal zone	Divisional office	Division
			Chief	Location
			Assistant Chief	Sub-location
Catchment	Topography and drainage pattern		Headman (no formal standing)	Village
Farm	Land use		Farmer	Farm
Field	Management		Farmer	Field

Source: McDermott *et al.* (2002).

various issues are connected through loop, or influence, diagrams, which I shall discuss more when we talk about making diagnoses. Figure 1.2 gives a general example of a possible systems description for the emergence of food-borne diseases: poor people need two career families and a cheap food supply; wealthy people want fast food, but, having read epidemiological studies about the protective effects of vegetarian diets, they prefer fresh vegetables. These combine to create a system of global trade, consolidation and intensification that leads to increases in both diseases and resistance. Through marketing, efficiency measures and economies of scale, which keep prices down, these patterns tend to reinforce themselves. Unless some dramatic event intervenes – such as mad cow disease in the United Kingdom – these patterns are very resistant to change. This kind of picture begins to help us make sense of the overall 'patient' and to look for ways of resolving conflicting outcomes.

What we are exploring here are the beginnings of what might be called a physiology of eco-social systems. Indeed, some scholars have developed a model 'societal metabolism' (see the journal *Population and Environment*, November 2000 and January 2001). However, this is considerably more complex than organismal metabolism.

For instance, we can also ask: who is doing the things that cause the problems? Who is suffering the consequences? Who is benefiting? What will become clear during this exercise is both the importance of the multi-scalar view of our patient, and the fact that there is no single legitimate description of the patient – and therefore no single person to 'blame'. We all participate in the pathology of the systems

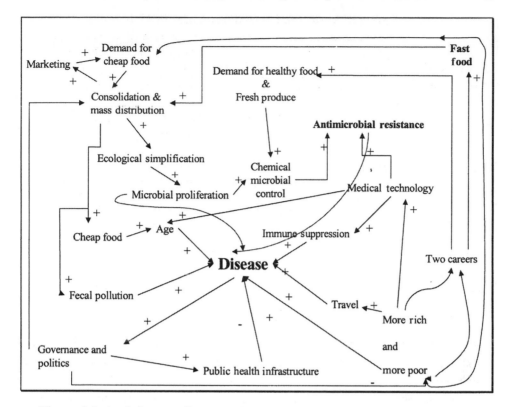

Figure 1.2 An influence diagram of causes of food-borne disease emergence (re-drawn and modified from Waltner-Toews and Lang, 2001).

in which we live. According to the Ecological Committee of the International Joint Commission, in the ecosystem approach 'there is not one material ecosystem to which our definitions must conform. Rather, the human actor must accept responsibility for erecting definitions and be prepared to change them when the purpose of the description changes.' (Allen *et al.*, 1993, p. 5).

Given that there can be many descriptions from many different perspectives, one challenge to creating system descriptions is that of selecting what to put into them and what to leave out. We cannot describe everything about everything! Nor does it seem appropriate for expert scientists (which scientists?) to determine what is important and desirable for everyone else. The scientific, ecological information is important, but not sufficient. Recognizing that there are many human actors, with many legitimate perspectives, Kay and Schneider (1994) have argued that using an ecosystem approach means 'changing in a fundamental way how we govern ourselves, how we design and operate our decision-making processes and institutions, and how we approach the business of environmental science and management'.

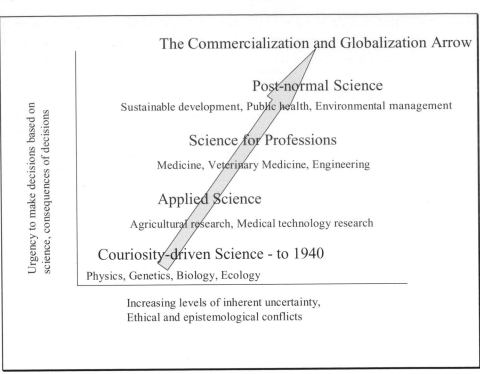

The Commercialization and Globalization Arrow

Post-normal Science

Sustainable development, Public health, Environmental management

Science for Professions

Medicine, Veterinary Medicine, Engineering

Applied Science

Agricultural research, Medical technology research

Couriosity-driven Science - to 1940

Physics, Genetics, Biology, Ecology

Urgency to make decisions based on science, consequences of decisions

Increasing levels of inherent uncertainty,
Ethical and epistemological conflicts

Figure 1.3 Science: from basic to post-normal (modified from Funtowicz and Ravetz, 1994).

Post normal Science

In the same vein, Jerry Ravetz and Silvio Funtowicz have developed the idea of extended peer groups and 'post-normal science'. In normal science, one's disciplinary peers determine the 'success' and 'quality' of one's work. If we are talking about sustainable, healthy communities, then clearly there are others who will have something important to contribute. This is especially important given the uncertainty of scientific predictions with regard to complex systems. In an often reprinted diagram of different kinds of sciences (Figure 1.3), Funtowicz and Ravetz situate 'pure science' in the lower left-hand corner, at the intersection of the x and y axes, where uncertainties are small, as are ethical and epistemological (knowledge) conflicts, and decision stakes. In the health professions, the conflicts, uncertainties and decision-making stakes increase; at the same time, the peer group expands. While an excellent post-mortem diagnosis might be viewed as scientifically acceptable, it is clearly viewed as a failure by this expanded peer group. When we are talking about such things as sustainable health and agriculture, all of these conflicts, uncertainties and problems are magnified by several orders of magnitude – and the peer group who will judge our success or failure also needs to expand. I have tried to situate various kinds of science and its applications on the original figure (see Figure 1.3), and have added an arrow, which does not appear in their work. In this way of looking at science, the expanded peer group is demanded by a combination of uncertainty in the science itself, and the uses to which it is put. I would argue strongly that the combination of commercialization and globalization has ravaged our 'innocent', 'objective', 'curiosity-driven' science. Commercial and political interests rarely allow us to learn about the world the way a child does, exploring and wondering. Basic biological insights into genetics now are immediately owned and commercialized. Amateur scientists and naturalists – and many aboriginal groups – are the last refuges of the kind of contextual, place-based, culturally rooted knowledge that we need in order to learn our way into a sustainable future.

One of the key steps in an ecosystem approach to health and agriculture is to find all those who have a legitimate stake in the problem we are studying and want to resolve. We will need to use some common sense here. Everything may be related to everything else, and everybody may have a stake in the environmental condition of the planet, but some problems will tend to accumulate locally (like toxic spills) while others (such as fossil fuel use) may tend to accumulate globally. As with everything else in life, we will need to make some ethical decisions, in this case to protect those who are most vulnerable to exploitation – or to work with them so that they can develop their own protections. In expanding our peer groups, we have to be very careful that we don't simply defer to those who are paying the money. Indeed, we may often have to resist the commercial–political alliances in order to regain our place on earth. The stories of those (people, species) who are paying with their health and their lives for current practices must be heard around the table.

Questions

Give a specific example to illustrate the clinical signs described in this chapter. Can you think of some other clinical signs? At what scale are those changes occurring? Describe the patient who is suffering those clinical signs by scale and perspective.

Take a particular presenting complaint as described, for instance, in a published paper, and classify it. Start with scale and add perspective. Do the broad social and ecological scales match? Try to combine your classifications into a matrix like that in Figure 1.1.

Begin sketching out some connections among clinical signs, based on your general knowledge and previous research. Now include some variables other than those related to the complaints, which you think might be important, but which people might think are good. How do the 'good' things relate to the 'bad' things?

2

The clinical examination: asking questions, getting data

Streams of inquiry

Once we have been presented by a mess of complaints, ranging from contaminated water to climatic change, and have begun to define our patient, how can we begin to make sense of it? How do we gather information so that we can make a diagnostic judgement and propose some solutions? What kinds of information do we need?

James Kay and his graduate students at the University of Waterloo, Canada, have proposed what has been called the 'diamond schematic' as a way of organizing our thinking about the kinds of information we need. Figure 2.1 shows one of several versions of this. It has been applied in a wide variety of settings – from environmental planning in Saskatchewan to adaptive watershed management in India – and appears to be both useful and robust.

The diamond schematic highlights the kind of information that is needed to effectively undertake an ecosystem approach, and suggests how that information can be organized or presented. In this chapter, I want to focus mainly on the top of the diagram, the two squares which come together to form a description of the eco-social system.

Describing an eco-social system involves both a scientific description of the ecosystem (plants, animals, soils, water, and how they relate to each other and to flows of energy and nutrients), and the creation of an 'issues framework' (which things are deemed important – either positively or negatively – by the people who live there), based on an understanding of the culture and values of the people who live in the system. Furthermore, these two sides of the description interact with each other; people's cultures and values will influence how the ecosystem is described, and vice versa.

The ecosystem description helps define what the constraints and opportunities are (the ecological possibilities), and the issues framework leads to the creation of visions and preferences. What do people want to see on this landscape?

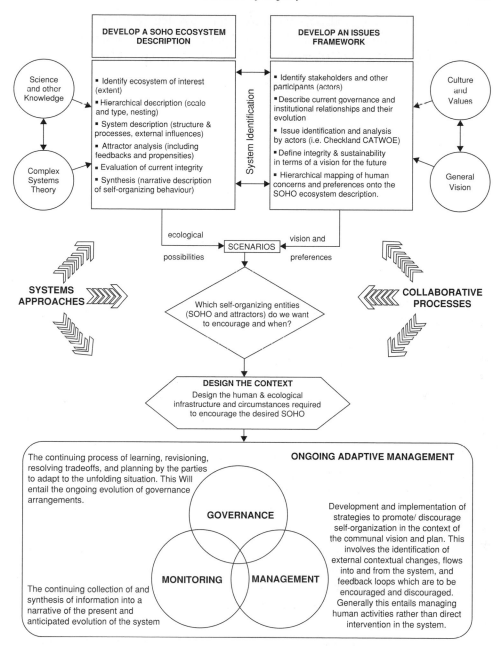

Figure 2.1 The 'diamond schematic', developed by James Kay and co-workers. Streams of inquiry for assessing and managing ecosystems.

All this sounds very good. So, we walk into an ecosystem or a community and . . . do what? If we look carefully at where we want to end up – a description of how social and ecological phenomena interact over time to create patterns of activities and outcomes – it is quickly clear that there is no single set of methods that will get us there. In fact, we will need to use a variety of methods.

We will need a good history of the place – including all the usual temporal parameters of epidemiological interest, such as diurnal, seasonal, periodic and secular patterns. Furthermore, we want in particular to see how socio-economic and political and ecological variables have interacted over various time frames, not just charts of particular outcomes. At the global level, the works of scholars such as A. J. (Tony) McMichael and Jared Diamond – writing in the excellent tradition of Dubos, Zinsser, Burnet and White, and W. H. McNeill – are essential to begin to understand our present health predicaments at the global level.

For specific projects or investigations, much of our information for the history comes from secondary data – both surveillance and monitoring types of data and socio-demographic data, often collected by government agencies. This information can be difficult to get, as various government workers try to protect themselves against public exposure. In general, governments have little incentive to explore problematic situations. They seldom look good and solutions often require financial outlays. Nevertheless, as scientists and citizens, it is essential that we pursue all forms of inquiry into where we live, and while the information we get this way may be incomplete or biased, it can serve as a useful beginning to our broader and deeper inquiries.

We will also want good political, economic and stakeholder analyses. Who has lived in this place? Who has made the important decisions? How have these interacted with natural changes in floods, rainfall, and seasonal and long-term droughts? To describe what is there now, we may need laboratory science and environmental field studies, as well as aerial photographs, secondary data analysis, and ethnographic and participatory research. What we are headed for is not some transcendent 'transdisciplinarity', but what has been called 'triangulation', evaluating a situation by shining many flashlights on it from different angles (Roe, 1998). If this sounds data intensive, it is. If it sounds impossible, be assured it is not. We gather information as we need it to answer particular questions, make decisions, and learn from our decisions.

Disease care practitioners – indeed applied scientists of all sorts – will recognize this way of doing things. You take laboratory tests, physical examinations and personal stories and combine them in some way to arrive at a rounded picture and, perhaps, a diagnosis. Some systems practitioners would like to reduce the entire process to one of creating many different diagrams and models. Like the myriad of systems models for human health, however, these are both instructive and problematic. The structured nature of the many models seems to suggest that one could computerize the process, yet most attempts at doing so have been less than wildly successful. Many of the models that are most richly descriptive are more like pictures, stories and myths than like mathematical models. Like clinical medicine and health promotion, practical ecosystem health work will always involve a large dose of clinical judgement. In this case, however, as we shall see later, judgement is not made by individual experts or authorities; that judgement is made collectively by people who care about the situation and know something about it. The expertise is *collective* rather than individual. This is another characteristic of ecosystem health work that distinguishes it from much conventional medical practice.

In the sections below, I outline briefly a few of the methods most frequently used in ecosystem health studies. Emery Roe has argued that, in triangulating, you want methods that are as orthogonal as possible. The best way to understand complexity is to focus lights on it from very diverse angles. Thus, in ecosystem health work, participatory action research, quantitative epidemiological studies, systems studies and landscape ecology all have their necessary places. As an epidemiologist, I find it useful to think of them in terms of the standard 'who, what, where, when, and why'. What we need to keep in mind as we go through this exercise, however, is that we are not simply after the kinds of linear or flat descriptions of a problematic situation that are the usual result of standard scientific investigations. We are in fact using traditional methods of investigation to arrive at a description of the system by scale and perspective, an understanding of who the owners of various issues are, and why the system behaves as it does. In all this work then, we must keep 'systems', and, more specifically, 'SOHO systems', in the back of our collective mind.

Starting with the problems (who, what, when)

One of the veterinary epidemiologists on the ecosystem health elective we run for Canadian veterinary students says that the best place to start talking about ecosystem health is next to 'a pile of deads'. He's exaggerating a bit. His point is that, for health professionals, the presenting complaint in an ecosystem health mess often has something to do with diseased or dead people or other animals. A reasonable way to begin investigating the mess, then, is to undertake an epidemiological study. In acute situations, this will take the form of an outbreak investigation: animals are dying, people are sick, the fields are washing away – why? This is not the place to explain the details of how to be an epidemiologist.[1] However, it is worth reviewing

---

[1] Many offline and some online resources are available (see http://www.bmj.com/epidem and http://www.pitt. edu/~super1/ for starters).

Table 2.1 *A classification scheme for sick and not sick
things in an epidemiological investigation*

	Sick (case)	Health (control)	Total
Exposed	A	B	A+B
Not exposed	C	D	C+D
Total	A+C	B+D	N

the basic structure of scientific field investigations and then seeing where we have to go to begin understanding ecosystemic approaches.

In experimental sciences, researchers take animals that they believe are alike in all respects and then randomly (not haphazardly, but according to strict rules of probability) assign them to getting exposed, or not exposed, to, let's say, a certain parasite. Then, if the exposed animals develop disease more often than those not so exposed, they conclude that whatever they were studying (the parasite) caused the disease (a cyst). In this setting, all other things being equal, we can be pretty sure about the causal links. However, in the real world, all other things are never equal. Furthermore, if the cause of disease we are interested in is a bad habit, like smoking, or a behavioural quirk, like dog ownership, we can't randomly assign them to some people and not others. Epidemiologists have recognized this and have developed various techniques for simulating experimental assumptions in the real world. The basic structure of classifying data, however, is still very much as in experimental sciences. The basic form is shown in Table 2.1, but it can be expanded to include many gradations of health and disease, as well as many gradations of exposure, ultimately shifting to continuous variables in either case.

For instance, in an investigation in Kathmandu of echinococcosis, a tapeworm of minor significance in canids that can create tumour-like cysts in other animals, including people, we took a random sample of households (n), and cross-classified them according to various characteristics (variables). For instance, did they or did they not allow dogs to defecate in the house (exposed or not) and did they or did they not have a lot of sick people in the house. This cross-sectional study gave us a snapshot of the neighbourhood we were studying. This 'two-by-two' cross-classified view of reality can be made complicated (more realistic) in a variety of ways. For instance, we can look at graduated exposures and graduated outcomes. A dose-response curve, such as that used in risk assessments, may tell us if low-dose exposures to some chemicals cause less damage than higher doses. Data from such studies would be presented as a line on an x–y axis graph. In population studies, we may examine if exposures to small amounts result in lower probabilities of disease than exposures to higher amounts.

There are some practical problems in doing a basic observational study as part of ecosystem health work. The first of these is sampling. You need a sample that is representative of the population you want to make inferences about. Usually this is done formally with a random probability sample. How do you take a random sample of ducks, or raccoons, or benthic invertebrates, or trees? Is there such a thing as a population of ecosystems? Even if there is, is it feasible to sample them and study them?

The denominator in a rate is easy to determine in a community (number of houses – or people, depending on what we're interested in) or in a barn (number of cows). But in a wildlife study, say you find a hundred dead ducks along a shallow shoreline. Is that 100 out of 1000 or 100 out of a million? Working with a zoologist or fisheries biologist who works with wildlife or fish is usually the best way to begin to get a handle on these kinds of issues. Wildlife veterinarian Gary Wobeser has also described the details of how one might begin to investigate these kinds of problems in his book *Investigation and Management of Disease in Wild Animals* (Wobeser, 1994). Dominique Charron, who studied benthic invertebrates and fish as indicators of stream health, has discussed some of the ways to obtain these data in her thesis (Charron, 2001).

Having established a rate, we still need to decide whether this is a normal death rate or not. Maybe a hundred ducks have been dying on this shoreline every year – or every third year – for the past two millennia. In other words, we need good long-term histories.

These problems are solvable if we work in teams of people with a variety of skills and knowledge. From ecologists, zoologists and wildlife disease experts we get techniques for sampling the wilderness. From environmental historians and ecologists we get a history of our patient.

Even if all these sampling problems are solved, there is another problem with cross-sectional studies. Remember that in all these studies we are attempting not just to document what is there at a particular moment, but to discern patterns over time. If we find that houses with high infection rates also have dogs which defecate in the house, how can we be sure which came first, dog pooping habits or disease? Do shallow shorelines cause duck die-offs or do dying ducks congregate on shallow shores? Does DDT accumulate in the shells of sick birds, or does the DDT cause the illness? Do viruses kill seals, or do they invade seals that are already sick for other reasons? Much of this we can begin to decipher by knowing natural history, pulling in laboratory data, and from experience with many similar situations. There are also some specific kinds of epidemiological studies that are helpful.

We can start with sick or infected people and compare them with healthy or non-infected people living in the same neighbourhood and roughly the same age. This is comparable to looking at micro-habitats for wildlife. What is it that differentiates people with the parasite from those without the parasite? Dog ownership? Living in a house where dogs defecated? Occupation? Economic status? Sex? What are the characteristics of the vegetation and water where the animals died versus those where they are healthy? In brief, we can come up with a set of possible risk factors, compare cases and controls, and determine if any of these allow us to discriminate between them. This is called a case-control study and is often used when diseases are rare, or in outbreak situations. For instance, this would be a good way to determine if people who reported diarrhoea to their doctors in Walkerton, Ontario, in the summer of 2000, were more likely to have consumed water from the city system than those who did not report illness.

Case-control studies are fine if the exposure (eating a contaminated food) closely precedes the effect. The initial research into mad cow disease (bovine spongiform encephalopathy) took this approach, comparing farms where the disease occurred versus those where it didn't. Given the late onset of the disease, many farms may have been misclassified as free of the disease when in fact they had it, but as a first approximation, the case-control study type served its purposes well. In the duck die-off, we might compare the levels of botulism toxin in the ducks along the shoreline with ducks that have died elsewhere.

In fact, pathologists have long experience with what is considered normal, and/or non-life-threatening, so they can often tell you from a direct single examination whether the toxin levels – or some surrogate like the number of maggots in the duck's stomach – are likely to be life-threatening. It will usually be pretty obvious whether the toxin came first, or the disease. It will still not be obvious whether the shallow shoreline came first, however. Case-control studies aren't much good for time-ordering.

If we were talking about Creutzfeldt-Jakob disease in people, or about some chemical exposure and its possible connection to breast cancer, where the delay between exposure and outcome may be a decade or more, we would certainly have trouble with the time ordering of the causes and effects. What did you eat ten years ago? How often did you eat it? And what if the characteristics we have chosen are not important? If exposure is through general environmental contamination in the neighbourhood, or through staple foods that everybody eats, then there will be no measurable differences between sick people or animals with regard to possible causes. A case-control study in Walkerton has difficulty identifying which well was contaminated, since all the wells contributed to the same distribution system. In fact, if everybody is exposed, then the cause will look personal, or genetic, since there is some inherent variability in resistance to every disease. If everybody ate *Salmonella* on a daily basis (everybody is in the 'exposed' category), some people would get sick and others not, the difference being attributable to personal habits or genetics. This is one reason why we should all be very suspicious of studies that claim to demonstrate definitive personal causes for diseases we suspect of being caused by environmental exposures. Most North American research is designed to NOT detect those effects. Since almost all population and ecosystem problems expose everyone within the system in different ways through feedback loops, we are dealing with what is called 'mutual causality' and standard methods of investigation and analysis are very weak and often completely inappropriate. We need to consider multiple sources of information and study types; in Walkerton, an engineering simulation model of the city system, combined with landscape and microbiological information, was able to identify Well 5 as the probable source. Nevertheless, most of us do begin by looking at problems in individual animals or people, where those methods are appropriate. The big problem is that most investigators end there, before they have even begun to unravel the true systemic nature of the problem. But I digress.

There are ways of getting around the time-ordering issue. If, in Kathmandu, we had strong prior suspicion that household dogs were the culprits, we could have found houses that were disease free, and following them over some period of time (say a year) compared infection rates in houses with dogs against those without. *"cohort study"* This is called a cohort study. The temporal stream is clear here, but we may run into problems if people move away, sell their dogs, acquire dogs, and so on. There are statistical ways to try to deal with these problems, but these problems are minor when put into the overall context of real life in a city like Kathmandu. What happens if we pick the wrong exposure and follow a hundred families for five years? That's a very costly bit of research for a negative result. From a research point of view, this may not seem like a major problem, since we may have ruled out a particular cause. It doesn't much help the current generation of citizens of Kathmandu, however.

Sometimes we can look back at say, wetlands, and pretend that we followed them over a decade or so – assuming that we can get that historical information. Old aerial photographs are sometimes used for landscapes, and hospital records for people. Time series analyses – different ways of plotting and analysing changes in probable causes and outcomes over time – are helpful to make sense of these kinds of data.

The above approaches do not assume any complicated connections between a web of causes that may work together or inhibit each other and a particular outcome. Epidemiologists, particularly those studying outcomes like cancer or heart disease, have recognized this and developed very sophisticated design and analytic techniques to begin to account for multiple possible causes. We can, for instance, study multiple connections in such a way that we can create path models, which look like overturned trees. We can build into these models the rate of disease transmission from butchers to dogs and from dogs to people. This enables us to stretch the chain of causation backward in time.

We can also try out, statistically, various combinations of alleged causes. Which are necessary? Which are sufficient? For instance, maybe only people in those households that had infected dogs who defecated in the house and had no running water got infected. For ecosystem health studies, where context is (almost) everything, these kinds of interaction terms are likely to be the rule, rather than the exception. Statistically, this creates really messy models, and practical interpretations are a problem. Is drinking wine, eating cheese and smoking worse than drinking wine, smoking and eating peanut butter? Is a farming system with cattle, pigs and rotating pasture management less stressful on the ecosystem than one with pigs, chickens and crops? Regardless of the impressions given by standard epidemiological textbooks, understanding even the rudiments of the epidemiology of even the most simple diseases requires us to draw on a range of qualitative and quantitative studies, at scales ranging from single organisms to populations of organisms – and to complex systems of which they are an evolving part.

The information generated in multivariable epidemiologic studies begins to get useful for ecosystem health studies, since we can include various kinds of economic, biological and behavioural causes in our picture.

None of these techniques, however, assumes any boundaries around the problem. The problem begins with a cause, or causes, and ends with an effect, or effects. The cause–effect connections are assumed to be universal. If tobacco smoke causes cancer, then this must be true in Kathmandu as well as New York. Once we identify the cause we can suggest ways to remove it.

Universally adopted linear cause–effect models seem to work pretty well for some high-profile agents within a short time-frame. However, they do not account for situations in which the outcome itself may, through some circuitous route, change its own causes. They do not, in other words, allow us to detect attractor states in which diseases and causes may be embedded. What if we suggest that people should treat dogs with a certain drug? This may mean diversion of some money from the general household resources to dog treatment. This, in turn, may mean that there is less money for food, in which case people will be less well-nourished and more susceptible to disease. Or maybe if the dogs are healthier, then the people will be healthier as well, since there will be fewer parasites transmitted across the species line. We can't be sure about what will happen in this regard. What if treating the dogs means more dogs survive, which means greater environmental contamination in the neighbourhood with more dog faeces? What if the drug passes through in the faeces of the dogs and not only kills the parasite, but also the insect larvae born from eggs laid in the faeces; and what if those insects are important for pollination of flowering plants which are an important source of income to those families, perhaps providing the extra money needed to buy the drugs to treat the dogs? Maybe small-holder beef cow-calf farms with pastures are good for the environment if they are organized in certain systemic ways, but bad if they are organized differently.

I think you can see where this is leading. Once you put some boundaries, however loose and ethically problematic, or layered (as in a holonocracy), around a problem, so that one or more of the outcomes affect one or more causes of the same (or other) diseases, the standard epidemiological techniques begin to weaken. If our unit of analysis is the ecosystem and the brains of the ecosystem are the geographic community, and all are nested, how does one begin to determine appropriate sample sizes? What are the comparison groups? What are the possibilities? What does 'cause' mean? In most situations with which we are faced – such as parasites or bacteria in water supplies, pesticides in the environment, or epidemics of malaria – the interactions between economic, health and environmental outcomes cross even more temporal, spatial and disciplinary boundaries with apparent disregard for scientific niceties. Events today in the tropics (DDT spraying for malaria control) influence outcomes ten years from now in the Arctic (pesticide residues in breast milk). Farmers' expectations of future weather determine which crops they plant today and probably influence the shape of food- and water-borne disease epidemics in the medium term.

In urban communities in Kathmandu, the picture of hydatid disease is complicated by shortages of fuel and water (constraints set within a larger Nepalese ecosystem), which are used to solve some public health problems, and whose heavy use in the city may create greater environmental problems in the countryside, more migration to the city and, within a generation, greater disease problems in the city. In Uganda, cattle can carry the trypanosome parasite, which causes sleeping sickness in people; the parasite is transmitted by tsetse flies, some of which like shady areas by streams. But if you clear out the bush near the streams, you could end up with serious erosion, and cattle are both nutritionally and culturally important (and ecologically important in maintaining the landscape). So do you just treat everything with drugs? We already know where that leads as an ultimate solution (it's good in the short term, but we need something long term in our back pockets).

If we have a disease outcome in which the incubation period is long, and in which infected people become a source of infection for insects who infect other people, and in which actual disease is only manifest if there are some nutrient deficiencies (none of this is far-fetched), then the problems of identifying useful causes – those which can be used to control disease without exacerbating a whole lot of other problems – becomes even greater. As noted earlier, Russel Ackhoff, a well-known systems scientist, refers to these kinds of sets of interacting problems as a 'mess'. A variety of systems inquiry and intervention methods have been developed to make sense of such messes.

Getting the picture (where)

In recent years, there has been renewed interest in the spatial distribution of health and disease outcomes, and with it, a renewed interest in techniques developed by geographers looking at spatial units of inquiry. Basically, these techniques – ranging from simple hand-drawn maps to complex overlays in computerized Geographic Information Systems (GIS) – are ways of creating visual pictures of the contexts in which health and environmental outcomes interact. This is a way of drawing physical boundaries around the problems we are looking at.

International attention is often focused on the 'sexy' colour-enhanced maps produced by GIS. However, spatial drawings may be drawn during workshops with local communities to help them set out not only what they see as being important, but how issues are connected.

Some maps are simple. Erin Sifton and Anita Beaudette, in an agro-ecosystem health study in Honduras, had villagers draw maps of their areas. Looking at the separate maps drawn by men and women was very informative. Men tended to draw greater detail around the village, and less detail inside the village, reflecting their daily activities. One area within the village was labelled a soccer field by the men, and a pasture for small livestock by the women.

Such mapping can become quite complicated. Rich Pictures, which have become a standard part of Soft Systems Methodology (see below), are drawings of a problematic situation usually drawn by researchers together with various actors as part of participatory data gathering activities. Martin Bunch used this approach at a series of Adaptive Environmental Assessment workshops in Chennai, India, by asking participants to draw pictures of what they thought were the important elements in the Cooum River valley, and how they might be connected. The drawing was first made on a large whiteboard, and later transferred to paper; modifications were made as the workshops progressed (Figure 2.2).[2] This picture was subsequently used as a basis for developing dynamic spatial GIS models of the area.

At another level of detail and sophistication, the Huron Natural Area project in Kitchener, Ontario (see www.jameskay.ca) made remarkable use of available

---

[2] Martin Bunch describes this in some detail in his PhD thesis. This interesting thesis was published in book form by the Geography Department at University of Waterloo, but the quickest way to access this is to go to the website of James Kay (www.jameskay.ca) and then search under 'various work of my students'.

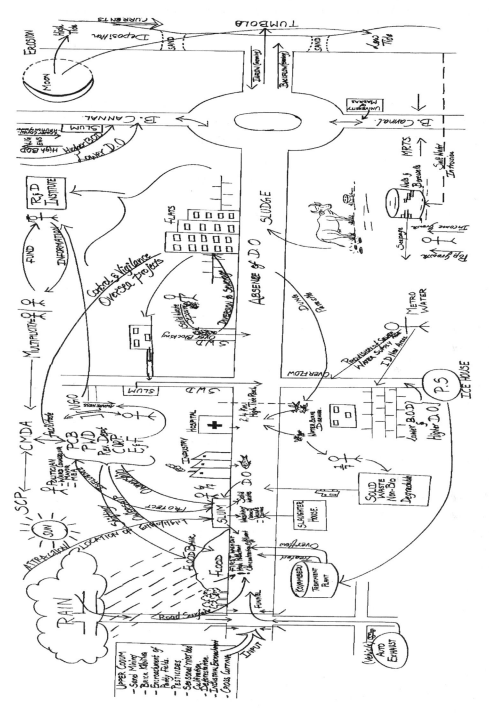

Figure 2.2 A Rich Picture of the Cooum River in Chennai, India (Bunch, 2000).

photographic data. Aerial and ground photographs demonstrated how the area had changed over time. Much to some people's surprise, this 'natural' area was shown to be abandoned farmland; much of the standing and running water in the area was shown to come after the trees arrived, and not before.

If enough is known about the social, climatic and ecological conditions under which diseases emerge, and the information is available to map these conditions, then sophisticated GIS techniques offer some hope of being able to pinpoint high-risk or vulnerable areas. Mapping rainfall variations and vegetation changes over time, using satellite images, for instance, has been used to predict the likelihood of Rift Valley Fever epidemics in Kenya (Linthicum *et al.*, 1999). This may be especially important when studying the possible effects of climatic events such as the El Niño/ La Niña Southern Oscillation Phenomenon or global warming. Drawing spatial boundaries around problematic messes is important for administrative reasons (who's responsible for this mess?). They are also important because carefully drawn spatial boundaries may represent ecological and social boundaries around interacting issues. They help define the skin of the patient we are looking at.

Ecological studies

It may seem odd, in a book that purports to take an ecosystem approach, to devote so little time to methods of ecological investigation. This is because these methods are well established in standard ecological texts, and because I expect health practitioners to work with ecosystem specialists in resolving ecosystem health problems. The information from ecological studies – ranging from a cataloguing of species present, to water, energy and nutrient flows, to trophic food webs, and normal ecosystemic changes over time – are essential for defining problems and seeking solutions. They provide the basis for defining constraints and opportunities for 'improvement'. Anyone attempting to promote ecosystem health should certainly be drawing on this information however it may be gathered – from mapping to collecting of field specimens, from talking with local naturalists and farmers to tagging animals and monitoring species. When we begin to draw systems diagrams to understand the data we have collected – and not merely to document 'problems' – this ecological information will be crucial. Furthermore, since social and ecological systems are now integrated as single eco-social systems pretty well everywhere, the application of ecological concepts to the study of social phenomena is an area that needs greater emphasis as we develop new methods of investigation and synthesis. Industrial ecology and the conversion of social interchanges into ecological currency are the cutting edge of this work. The web page of James Kay (www.jameskay.ca) and the work of Mario Giampietro and colleagues (2000, 2001) are full of examples of this kind of work.

Participatory methods (who and why)

Although we can glean a great deal of information from secondary data sources, and by collecting 'hard' data in the field, good scholarly inquiry requires us to actually talk to people. While scientific investigations often take on themselves a mantle of 'objectivity', they have often been shown to reflect strong biases. Not all information has been written down, and, if it has, it may have been written down in biased ways, or selectively. In many ways, getting the technical information that most scientists and professionals are comfortable with (number of ticks, varieties of plant species, water flows) is the easy part of an ecosystem approach. As any good health worker knows, however, it is absolutely essential to talk to the complainants (sick people, animal owners) in order to get historical information and a sense of why certain outcomes might be viewed as negative and others as positive, as well as to get a sense as to what might be feasible courses of action to resolve the issues.

When a person is sick, we talk to that person, or, in some cases, a nearest relative, loved one or guardian. For a sick animal, we speak to the owner. When something is wrong at the farm or household level, things already get complicated. A lot of agricultural and health development programs have fallen flat on their faces because those who want to 'help' are talking to the wrong people or not enough of the right people. Animal husbandry programs, for instance, have been designed based on interviews with male farmers in situations where women have been the primary livestock managers. Often, children are primary animal caretakers. In a household where there is abuse by the household head, that person is probably the wrong one to consult about creating healthy families. (Or at least, that person is not the only one who should have input.)

Beyond the farm and household, when we get to the community, which is the primary social locus for ecosystem health studies, the complexity rises exponentially. How do we define the community? What are its boundaries? Who is the 'owner' of the problems? Of the solutions? There are no simple formulas for this. What if the men and women differ not only in their diagnosis, but also in what they consider to be 'facts'? If the men legally own the cows, but the women get all the money for the milk, and the men get the money from the sale of the cow, but can't sell it without the woman's permission, who really owns the cow? Ricardo Ramirez has devised a framework for undertaking a stakeholder analysis and conflict management in the context of natural resource use. Figure 2.3 sets out this framework in terms of propositions. The first five propositions deal with identifying the stakeholders and describing them; the last four propositions deal with conflict management after

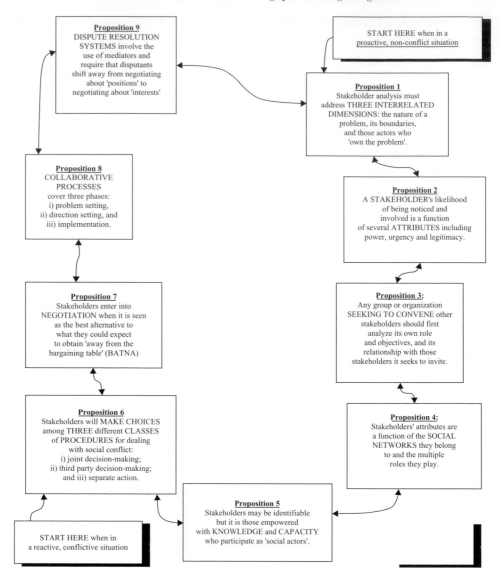

Figure 2.3 A conceptual framework for stakeholder analysis and conflict management (Ramirez, 1999).

stakeholders have been identified and they are trying (or not trying, as the case may be) to set goals for the system (Ramirez, 1999). What is important is that those who convene the stakeholders be aware of their roles and power in the situation.

There are a lot of ways of doing 'participatory' work in communities. Some people might even argue that when you go and ask a farmer some questions, she is 'participating' in your research. Certainly, semi-structured interviews are a form of participation. While we can often use information that we get through interviews and surveys, we generally mean something different (more) when we talk about Participatory Action Research. Usually we mean that we are working together with the people who live somewhere (the owners, actors, customers in Soft Systems Methodology terms, or stakeholders, or the public – the people who actually have a *real interest* – in what happens) to come up with descriptions of the problems and strategies for solutions to those problems. Often (not always – this depends on the presenting complaint) we are working with groups of people – neighbourhoods, communities, villages, focus groups – rather than individual people or households. And often we have to work with sub-groups separately – women and men, landowners and field workers – before we start to bring them together.

How does one actually go about doing this kind of inquiry? A variety of books and materials are available, many of which are included in the reference list at the end of this book. The guidebook by Rennie and Singh (1996) is an excellent place to start. The book by Pretty *et al.* (1995) explains many of the tools that have proven useful, although it is important to be clear that these tools must be adapted to each situation, depending on the local cultural context and the questions you are trying to ask. *Health Research in Developing Countries* edited by Joyce Pickering (1997) also has some good summaries – as well as tools for evaluating the degree to which your PAR work is actually 'participatory'. On the web, the Resource Centres for Participatory and Learning Network (www.rcpla.org) provides an excellent door into this literature. The International Institute for Environment and Development (www.iied.org) also has some materials on their site. Many participatory methods involve picture-drawing and/or using objects to demonstrate elements in the system in order to by-pass literacy issues and access local knowledge more directly.

Some participatory methods are designed to educate people, some to mobilize them to action, some to get information from them. In an ecosystem approach to health, we want to do a bit of all of this. Thomas Gitau, in his work on agroecosystem health in six Kenyan highland villages (McDermott *et al.*, 2002), used many of the available tools, ranging from semi-structured interviews and workshops to community resource mapping and transect walks. He also worked with men and women separately, as well as together, in order to obtain a multi-ocular view of the situation. You can read more about this study, and others like it, in the Projects section of the NESH website (www.nesh.ca), as well as in Waltner-Toews *et al.* (2004a).

While not discussing participatory methods in detail, I am going to expand just a bit on two participatory methodologies because of their direct connections to systems sciences: Soft Systems Methodology or SSM, and Adaptive Environmental Assessment and Management.

Soft Systems Methodology

Both Soft Systems Methodology (SSM) and the Adaptive Environmental Assessment and Management (AEAM) can at one and the same time be viewed as methodologies for inquiry and management processes.

Peter Checkland has promoted a way of thinking about complex problems involving people that he has called Soft Systems Methodology, as distinct from hard systems methodology, which one might use to study, say, the ecology of cow manure. Geographer Barry Smit prefers to call them 'difficult systems' versus 'hard systems', terminology that more accurately reflects what we are talking about. In general, Checkland takes a manager's point of view and encourages us to think in terms of Human Activity Systems, that is, the systems we use to define and solve problems, rather than in terms of systems descriptions of the problems themselves. This approach has provided the underlying rationale for the Ecosystem Approach used by the International Joint Commission in looking for ways to manage the Great Lakes Basin sustainably, both environmentally and economically (see also Waltner-Toews *et al.*, 2004a).

The original SSM process was seen as having seven components (Figure 2.4). We begin, first of all, with a problem situation, rather than merely a problem. We further recognize that, for most social problems, which environmental and public health problems most assuredly are, we can identify various human activity systems that are contributing, in one way or another, to that problematic situation. You can see that we have already shifted our attention away from the real world where we (and all our applied professions, such as engineers, physicians and veterinarians) perceive the problem to be, to the human activities that have created the problems. For those of us interested in solving problems, and not just describing them, this approach looks hopeful.

Once a problematic situation is identified, it is described through the drawing of various Rich Pictures and other participatory visualizations. Then, the situation is described through a series of systems models, each having as its basis a root definition. Checkland uses the mnemonic CATWOE to help in remembering what kinds of things should go into a root definition: clients, actors, transformation (what the system does), *weltanschaung*, owners and environment (what is externalized) (Table 2.2). This definition gives the human activity system a name in such a way that we understand why it exists and what it is there for. It links a transformation process (from some input, like plants, to some output, like food) with a particular *weltanschaung*, or world-view, which makes the transformation meaningful in context.

Thus, a farm may be described as a system of people and machines to transform naturally growing plant material into food and feed suitable for human and other

Table 2.2 *The elements of a root definition*

C	Customers	The victims or beneficiaries of T (who is affected by this system?)
A	Actors	Who does T?
T	Transformation	The conversion of input into output
W	Weltanschaung	The world-view that gives T its meaning
O	Owners	Who owns the system? Who could stop it?
E	Environment	The context – what you take as a given

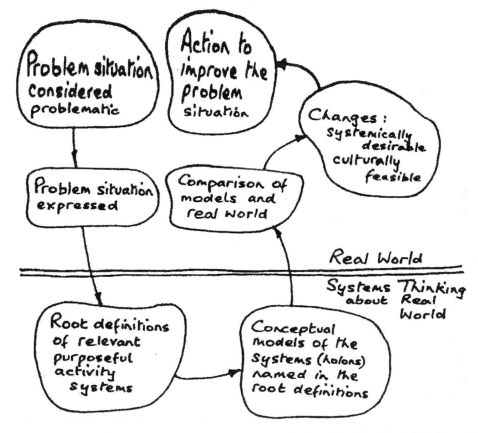

Figure 2.4 Soft Systems Methodology: the basic process (Checkland and Scholes, 1990).

animal consumption. It may also be described as a system for transforming naturally structured landscapes into human-structured landscapes, or as a system for transforming individuals into active members of a cultural community.

Once the models are developed, they are compared with the real-life, complex situation, and then desirable and feasible interventions are devised to improve the situation. This is a very logical way of working through problems as perceived by

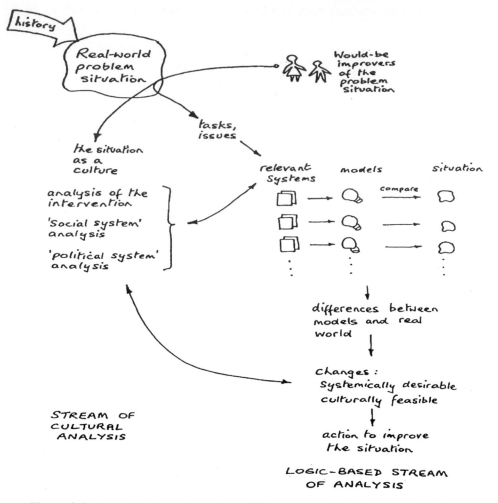

Figure 2.5 An enriched version of the SSM process (Checkland and Scholes, 1990).

human organizations. Checkland recognized, however, that viewing the problem logically was insufficient to understand why something was seen to be a problem. In later versions of SSM, he introduced processes of cultural (social and political) analysis to parallel the logical stream of analysis (Figure 2.5). Thus, several solutions might be systemically desirable (perhaps by promoting greater adaptability or stability), but not all of them would be culturally feasible. Anthropologists and sociologists work alongside scientists and community activists to help the participants understand why some things might or might not work for cultural or political reasons. Cultural and political contexts can be changed, of course, but require longer time frames and more complex discussions than simply working within existing organizational and cultural situations to assess systemic connections and negotiate trade-offs.

SSM, as I mentioned, can also be seen as a form of management. Certainly, the drawing of Rich Pictures and building of multiple conceptual models are ways of organizing information. However, these occur in the context of a process, which involves the actors (stakeholders, decision-makers). The 'managers' are thus part of the information-generating process. They are also part of the process of deciding on feasible and desirable goals. And, ultimately, they must be part of what substitutes for hypothesis-testing in SSM – making changes and seeing if the actual results agree with those expected based on the conceptual models used. This testing of the models and decisions by intervention is what we would normally call management.

Adaptive Environmental Assessment and Management (AEAM)

Adaptive Environmental Assessment and Management is an approach to managing natural resources developed in the 1970s by a working group at the International Institute for Applied Systems Analysis (IIASA). Carl Walters and C. S. (Buzz) Holling have probably been its most famous proponents.[3]

The AEAM process, as currently practised, shifts emphasis away from reliance on academic research, complex mathematical models and predictive planning to intensive workshops among managers, brainstorming, visioning and negotiating trade-offs. In its integration of research and management into a seamless process, it is thus similar to SSM. The original notion of AEAM suggested that the main aims of the process involved structured synthesis and analysis, development of predictive models, good monitoring, and the use of management activities to learn about the natural resources being managed. In the years since the ideas were first introduced, most scientists and managers have become considerably less optimistic about our ability to predict natural changes. However, the basic notion that we should treat management activities as experiments, and that therefore monitoring of changes and adaptation to them are important, has gained wide currency.

One of the important elements in AEAM has always been the use of workshops. Carl Walters, in his original book on the subject (now unfortunately out of print), said that he thought more was often accomplished in an intensive workshop of a couple of days than could ever be done through months (years?) of standard scholarly and managerial assessment and planning. I would tend to agree. A group of knowledgeable people working intensively – away from their usual places of work so they can't check their email or respond to urgent messages from their

---

[3] Good introductions to AEAM can be found in a variety of places. As well as the thesis by Martin Bunch mentioned earlier, the website of the British Columbia Department of Forestry is full of useful information (www.for.gov.bc.ca/hfp/amhome/introgd/appen1.htm).

bosses – can often quickly define the problems and come up with the best possible options for solutions.

Some researchers think this is easy, that doing a workshop just means getting a group of people in a room around a table, giving a few speeches, and coming up with some ideas. In fact, running a good workshop requires a lot of skill, so that everyone gets a chance to contribute, all ideas (no matter how crazy they seem) get on to the table, criteria are set up for grouping and/or winnowing items . . . basically so that it doesn't on the one hand degenerate into a chaotic mess, or, on the other, get stuck in the same old boring ideas and ways of doing things that everyone knows don't work but nobody can abandon because, well because that's how we do it, or that's policy, or that's what *I* think, and everybody knows I'm the expert. Cross-cultural work (how do you get a shaman and a scientist to cooperate?) is even more problematic. Many of the skills needed to run a good AEAM workshop are the same as those required to do good PAR work in general.

Bunch, in his workshops, drew on a series of documents published by the United Nations Centre for Human Settlements (Habitat) to structure one of his initial workshops. In particular, he used, as the basis for problem definition, the following series of questions from *A Guide for Managing Change for Urban Managers and Trainers* (Habitat, 1991; pp. 62–64).

> What is the problem? (start with a rough description; underline the key words and phrases).
>
> Why is it a problem? What would the problem look like if it were solved?
>
> Whose problem is it? Who owns it? (Once you determined who the problem belongs to, go back and underline all those you believe are willing to invest in its solution and, finally, circle the individual, group or organization you believe is the most important in the problem solving venture).
>
> Where is it a problem? Is it localized and isolated, or is it widespread and pervasive?
>
> When is it a problem? (e.g. every Monday morning at 8 a.m.; once in a full moon; only when it rains; when the boss is in town). As with other questions, be as specific as possible in your answer.
>
> How long has it been a problem? If it is a long-standing problem, this may say something about the ability, will or priority to solve it.
>
> Really now, what is the problem? Go back to your statement in Step 1 and determine whether: (a) the problem you defined is a symptom of a bigger problem; or (b) a solution to what you think is the problem. If you decide you are dealing with either symptoms or solutions, go back to Step 1 and try to identify the real problem.
>
> Finally, what would happen if nobody did anything to solve the problem?

Walters set out two tables, which contrast conventional approaches to environmental management with AEAM (Tables 2.3 and 2.4). I introduce them here as ways of gathering information, but, as with SSM and many other methodologies relevant

Table 2.3 *Conventional versus adaptive attitudes about objectives of formal policy analysis*

Conventional	Adaptive
1. Seek precise predictions	1a. Uncover range of possibilities
2. Build prediction from detailed understanding	2a. Predict from experience with aggregate responses
3. Promote scientific consensus	3a. Embrace alternatives
4. Minimize conflict among actors	4a. Highlight difficult trade-offs
5. Emphasize short-term objectives	5a. Promote long-term objectives
6. Presume certainty in seeking best action.	6a. Evaluate future feedback and learning.
7. Define best action from set of obvious alternatives	7a. Seek imaginative new options
8. Seek productive equilibrium	8a. Expect and profit from change

Source: Walters (1986, table 11.1).

Table 2.4 *Conventional versus adaptive tactics for policy development and presentation*

Conventional	Adaptive
1. Committee meetings and hearings	1a. Structured workshops
2. Technical reports and papers	2a. Slide shows and computer games
3. Detailed facts and figures to back arguments	3a. Compressed verbal and visual arguments
4. Exhaustive presentation of quantitative options	4a. Definition of a few strategic alternatives
5. Dispassionate view	5a. Personal enthusiasm
6. Pretence of superior knowledge or insight	6a. Invitation to and assistance with alternative assessments

Source: Walters (1986, table 11.2).

to practical ecosystem health, the boundaries blur between gathering information, making diagnoses, acting on them and assessing them.

Many of the attitudes and tactics set out here are reflected in the work reported in the websites referred to above. The main thing to note, I think, is that the new scientific attitude is based on a sense of fundamental uncertainty about our ability to predict, and therefore requires some humility and openness to new options. How this differs from, say, management work-shopping, or participatory action research *per se* is that it still remains grounded in an explicitly scientific view of the world. The peer group and the methods of investigation are not those of structured laboratory experiments. Nevertheless, we still seek to structure our activities (policies, management) in such a way that we can learn from them and adapt to new situations and new knowledge. This is what differentiates participatory action research from 'pure' political action.

The AEAM process and management workshops are very useful for getting managers to think carefully through the issues they are dealing with, and to focus on a particular problem. This raises two important questions. How does a managerial workshop relate to a community-empowerment workshop? Do the managers see themselves as servants of the community, and consult regularly with the community? Or do they see themselves in the role of Plato's 'philosopher kings' or America's CEOs? Secondly, many systems scholars would argue that what we are faced with is not 'a problem' to solve, but a whole mess of interacting issues, some of which are problems and some solutions – and some both, depending on your perspective. Indeed, even in Chennai, where Martin Bunch worked, squatter settlements along the river could be seen as a big problem by the watershed managers, and as a solution to lack of affordable housing by the squatters themselves. Checkland prefers the term 'problematic situation' to describe such messes. Stafford Beer speaks of 'problem jostling' and Russel Ackhoff, who actually uses the term mess for this complexity, speaks of problems dissolving as part of the systemic intervention and management process (cited in Flood, 1999).

These methods, then, if used as 'stand alones', may work best where there are clearly definable organizations to manage (farms, hospitals) and/or identified managers. They may also play an important role as *part* of the pluralistic mix for ecosystem health management, but you may need to spend a few years actually building up social organizations before you get anything done.

Investigating the non-problems

Except for basic ecological studies, the methods of inquiry covered in the preceding sections are based on solving problems or resolving problematic situations; furthermore, they tend to focus our attention on the problems themselves. In many ways, as I shall discuss in a later chapter, this is more like a medical than a health approach. In order to make improvements, however, it is important to understand the strengths and opportunities – the assets – inherent in a system, so that stakeholders can build on those. Otherwise the whole process of inquiry becomes one of medicalization and victimization. There are those who would argue that medicine (at individual levels) and epidemiology (at population levels) have done exactly that: identified everything that is wrong, little that is right, and in so doing fostered dependency and ill-health.

If ecological studies present us with a less problem-focused view of natural processes, and one that must serve as the basis for any future plans, one school of management refocuses attention on social assets. Appreciative Inquiry (AI) was

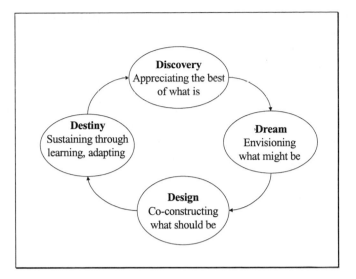

Figure 2.6 The Appreciative Inquiry cycle (adapted from the IISD website).

developed in the early 1990s by David Cooperrider at Case Western Reserve University. Its purpose, like that of many other systems intervention approaches, was to foster competitive advantage in business. Since then, AI has been extended and applied to community development, where it is used to build on achievements and inspire positive change.[4] In many ways, the AI cycle (Figure 2.6) is similar to approaches such as SSM, except for its more positive 'spin' on development.

A colleague from Nepal recently wrote to me and said AI was important because it built on the long and rich cultural history, capacities, knowledge and skills which people already have. It doesn't create the illusion that an outside expert knows better and is trying to fix something broken. In the sustainability project, we are all working together, drawing on each other's best instincts and skills, to learn our way into a healthier future.

Beyond management participation

SSM, AI and AEAM were designed for use by managers and workers in companies. These methods thus tend to assume corporate goals, to which individual activities are subservient. This works fine for corporations and farms. In the complex holonocracy of communities and ecosystems, however, these corporations become

---

[4] The International Institute for Sustainable Development (IISD) has an excellent discussion about AI on its website (http://iisd1.iisd.ca/ai/default.htm) and there is a world-wide portal for discussion by practitioners at http://appreciativeinquiry.cwru.edu.

Table 2.5 *Participatory tools used in an agro-ecosystem health project in Kiambu District, Kenya*

Activities	Tools	Data to be captured
Day 1		
1. Introduction	Ice-breakers Self introduction Logistics (meals, groups/ teams) Social maps	– Develop rapport, social structure of the village
2. Knowing the village (geographical/ administrative units)	Resource maps	– Physical structure of village – Natural resource inventory – Land use patterns, problem identification
3. Historical background	Historical profile	– Historical background – Major events and their impact on community – Problem identification and coping strategies
4. Trend and time lines	Trend lines[a]	– Resource availability and distribution over time and space – Diseases and pests dynamics – Infrastructure
5. Seasonal activities and trends	Seasonal calendars	– Yearly schedules – Agricultural activities – Effects of climate on agriculture
6. Mapping out route for transect walk and evaluation	Maps	
Day 2		
	Transect walk	– Natural resource inventories – Topography, village structure, farming systems
7. a) Triangulation and field observation	Semi-structural interviews	– Land use, lay-out
b) Drawing the transect profile	Profile	– Sources, incomes, expenditure
c) Livelihoods	Livelihood analysis	– Sources of goods and services, and market quantities of goods and services
	Mobility chart	– Institutions and relations/ linkages – Roles/responsibilities

Table 2.5 *(cont.)*

Day 3		
	Venn diagram (chapati)[b]	– Sources of information/ information flow – Problems related to institutions and linkages
8. Identification and analysis of institution	Information flow Map/charts	– Inventory of activities by gender and age Labour distribution
	Activity profile/ daily calendar	– Health concerns by age and gender – Impact on their productivity – Coping strategies
9. Health analysis	Health analysis	– Inventory of resource ownership, access and control by gender and age
10. Analysis of major gender concerns in AESH	Access and control profile/matrix	– Problems related to access and control of resources
Day 4		
	Decision-making matrix	– Causes and effects of the problems – List of major problems in order of priority
11. Problem identification and analysis	Scoring matrix Pairwise ranking	– Lists of opportunities – Means and ends of the opportunities
12. Needs identification and assessment	Problem tree	– Resources /inputs, – Responsibilities time-frame
Day 5		
	Scoring matrix Pairwise ranking	– Description of the problems, objectives, beneficiary community, detailed budget strategy of implementation
13. Action planning	Objectives tree CAP Proposal write-up	

Notes: [a] In a trend line, a group of older villagers draws one large graph showing, for instance, changes in population, disease rates, water availability – i.e. any variables they deem to be important. The vertical axis is qualitative (when a line goes up it means more); the horizontal axis is time, often marked off by key historical events (military coups, earthquakes, wars, famines).
[b] Villagers list all the organizations working in their village (government, non-government, unofficial, ad hoc, etc.). They then assign a large, medium or small 'chapatti' to each, according to how important they deem each to be. Finally, they lay them out on the ground and show how organizations cooperate or not on various activities by overlapping them to various degrees. These informal Venn diagrams, become, among other things, a means of identifying the organizations one should work with in order to achieve programmatic aims.
Source: Gitau *et al.* (2000).

one actor among many – and often these are actors who are creating problems as quickly as they solve them.

Participatory methods can also be used to educate people about where they live and empower them. Indeed, this is their main aim within the context of an ecosystem approach. Thus, they are more likely to draw on the philosophy of Paulo Freire than on the methods of Peter Checkland.

Many of the developments of participatory work within the ecosystem approach have involved adapting management tools such as those of AI, SSM and AEAM for use by people who live in dysfunctional eco-social systems and want to change them. Thus, while the overall process may look like SSM, and workshops may appear on the agenda, the tools used (see Table 2.5), the range of participants, the need for clear identification of trade-offs, rules for negotiation and conflict resolution and the like make it a substantively different – and perhaps more subversive – sort of activity.

Unlike many management programs, within the ecosystem approach effective local democracy is necessary not just to carry out programs but to set goals and create the programs. Also, unlike many social activist programs, such as those based on Freirian pedagogy, the ecosystem approach must incorporate broad public science, and education about ecological issues, as well as address issues of power and control.

Thus, local and indigenous knowledge has a place at the table in the ecosystem approach with externally derived scientific knowledge. Standard scholarly approaches are necessary, but insufficient. What have come to be called ethnoveterinary and ethno-medical and ethno-ecological knowledge are also necessary. No one kind of knowledge is taken purely at face value. Neither scientists nor local people are always right. This kind of 'public' or 'post-normal' science puts scientists in the uncomfortable position of having to explain and justify what they are saying to people with very different backgrounds, world-views and goals. It is a difficult task, but one well worth the journey, for, in the end, the sustainable health of people on this planet depends on our being able to convincingly tell our stories to each other.

Questions

Based on your list of complaints, and the scales at which they are apparent, describe how you will get the information to make sense of those complaints.

Do a stakeholder analysis, using Propositions 1–5 of Ramirez' framework. What is the community with which you need to work to resolve the issues you think are important?

Carry out a soft systems analysis, setting out the CATWOEs for each model. How many different versions of the system are there? Are they compatible? To what extent do the various communities you are working with redefine the issues you think are important?

What are the power relationships within the communities with which you are working? To what extent do they even comprise communities? Are some stakeholders more important than others? Why or why not? What are the implications of these power relationships and levels of importance for your investigation?

3

Making a diagnosis: synthesizing information from data

Making sense in a post-normal world

We have now heard both complaints and hopes for the future, and we have amassed a great deal of information. We have an idea of who the patient might be, although this can change over time as we learn more together with various stakeholders. However, in terms of the Basic Figure, we are still at the clinical exam stage, describing the system and identifying owners. Now we want to move on, to try to make some sense of it all. After all, we haven't been gathering data in this exercise simply to satisfy our general scientific curiosity. We are trying to sift through a range of issues to work with people to devise sustainable, healthy futures.

In many conventional scientific studies, what we are after is the ability to predict what will happen under particular circumstances, and then to either foster or alter those circumstances. In complex eco-social systems, we are faced with a dilemma. Commenting on the official Phillips report on the BSE epidemic in England, an editorial in the *New Scientist* commented that 'governments and the governed [must] become comfortable with notions of uncertainty and risk' (4 November 2000). In a series of publications, Silvio Funtowicz and Jerry Ravetz have elaborated a public, open, 'post-normal' science for sustainability (Funtowicz and Ravetz, 1993, 1994; Ravetz, 1999). A central task of this new science is to deal with the irreducible uncertainty inherent in eco-social complexity. Much of the literature on complex systems explains to us that we cannot actually predict what will happen in such systems.

But we have not lost predictability altogether. For many diseases, we have been able to create programs to eliminate them or reduce their incidence. We have a pretty good idea what happens in a watershed if you cut down all the trees. Even for complex systems as a whole, we can study how animals, plants, people and environments interact on a landscape in such a way as to create limited sets of patterns. We may not be able to describe linear cause–effect relationships in complex eco-social systems. We may, however, with philosopher of science Karl Popper and ecologist Robert Ulanowicz, describe propensities of such systems to behave in certain ways, almost as if their internal dynamics draw them into a limited set of possible futures, or propel them into certain trajectories. Complex systems researchers, such as Kay, refer to the set of options available to a particular system as its canon. We have available various visual, verbal and mathematical models which give us insights into that canon.

"canon"

If we cannot speak about predicting in any strict sense, what can we talk about? I think we can learn much about this kind of uncertainty by looking at how disease epidemics are investigated.

Every good outbreak investigation ends with a plausible story. This is not a statement of certainty that if someone neglects to turn a switch at the pasteurizer or doesn't wash their hands in the kitchen, people will get sick and die. This is a story that explains how, in this particular case, these people got sick and died because these actions unfolded in this particular way. Whether we are doing an outbreak investigation, a community-based ecosystem study, or a clinical exam on a sick person or animal, the story – the way we put the specific details of the eco-social history together – is what will determine how we should proceed. Like a cautionary tale, the outbreak story serves as a way to anticipate what is likely to happen under particular circumstances, and how we might avoid those circumstances in the future.

The story of how a cat got into the garbage when it was locked out one evening will tell you why she is sick today. Here is another story. In July, under a hot sun, several low-paid field workers in the southern United States were having bowel problems. They were paid so poorly, and the toilet facilities were so far away, off at the edge of the big field, that they could neither afford to take time off nor take a proper toilet break and wash afterwards. The distributor was looking for the best price on lettuce and he got it from that farm. The lettuces were then sent off to restaurants in the city. Several weeks later, a random assortment of people began reporting into doctor's offices with signs of hepatitis. Once we have the story, we can begin to think about what we will do differently tomorrow. Until we have that story, we're just messing around. In ecosystem health – as in evolutionary biology and all the health professions – the story is everything.

In order to promote ecosystem sustainability and health, we do not just want any story, but a collective story that can make sense of all the available information. Ninety-nine percent of the problem we have in learning our way out of the unsustainable mess we live in is that we are oblivious to the story we are living. We think life unfolds in some predetermined way, and thus we are unable to imagine an alternative way of living. However, this firm belief in predetermined paths simply reflects the stories we have internalized. These stories can be changed. We build our sense of self from stories (Bruner, 2002) and stories have been used, very effectively, to provide therapy for abused individuals and dysfunctional families (White and Epston, 1990).

So how do we get from a lot of 'facts' derived from different, orthogonal perspectives to a reasonable, collective story? As I said earlier, we may actually create word narratives. Or we may also use statistical or visual models of various sorts, which are also stories of a kind.

Standard statistical approaches and their limitations

After conducting epidemiological and ethnographic studies, there are standard sets of analytical tools that can be used to tease apart causal webs and identify problems. Many texts are available that deal with both qualitative and quantitative data, and I would refer you to standard statistical, qualitative data analysis and ethnographic data analysis texts for those. Basically what these tools do is to help us understand what kinds of things tend to occur in groups (are associated) – smoking and lung cancer, for instance, or poverty, health problems and joblessness. In recent years, some concerted efforts have been made to extend epidemiological techniques to study ecosystem-scale problems over time. McMichael (1993, 1999) gives examples of situations where income inequalities, loss of social capital, as well as air or water pollution can be or have been studied by stretching and adapting conventional epidemiological methods.

In Canada, several studies of water-borne diseases have combined spatial with temporal analyses to determine how water turbidity or agricultural land uses relate to human gastro-intestinal illnesses. A report on one such study in the Greater Vancouver area has been carried out by Health Canada (Aramini *et al.*, 2000).

What should be clear, however, from our discussions of linked ecosystem and health problems is that standard statistical and analytical methods become weak tools for understanding complex eco-social situations where various activities interact in ways that make the words 'cause' and 'effect' seem poor approximations of real life. It is important and useful to use standard scholarly and analytical approaches, if only to uncover subsets of problems that are easily managed – where to build toilets relative to water intake pipes, for instance, or what to do with organic waste. Once we have dealt with those, however, we can begin to talk about the kinds of communities and ecosystems we want to live in, the kinds of biological and cultural diversity that will enrich our lives.

Looking inside the boundaries: the idea of systems

One of the ways in which researchers have attempted to get their heads around complex problems, to tell a coherent story about apparently disparate elements caught

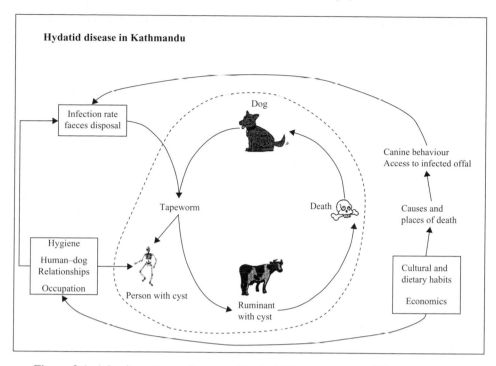

Hydatid disease in Kathmandu

Figure 3.1 A basic systems diagram of hydatid disease: externalizing context.

in a spaghetti-like mess, is by drawing boundaries around them and thinking about them in terms of systems, which were introduced in Chapter 1 and around which we have been skirting for the past few chapters. Systems are not stories, and some scholars see a conflict, or at least a tension, between systems as stable entities and systems as unfolding stories (Hayles, 2000). I see them as being necessary complements to each other. Systems studies provide the material, characters, relationships, the internal dynamics and the context for stories. Let's explore this a bit further.

For instance, we might draw a series of connections between dogs and people in Kathmandu (Figure 3.1). The area inside the circle is the system, and the area outside the circle is the environment. The various elements (dogs, people) are connected through different kinds of relationships, through which, by means of materials, information or energy, they communicate with each other and control each other.

Because of these relationships, the whole system, collectively, may respond to an outside stimulus in particular ways. Think of a dog, for instance, and how the dog might react if you stuck a needle into it (perhaps this will give you some empathy for veterinarians). All of the internal parts of the dog interact to respond, collectively, to that outside stress. This is one of the things that makes a dog a system; it is how you tell a dog apart from its environment. The sofa on which the dog is sitting doesn't bite you when you stick it with a needle. One could create a model of the nerve and chemical pathways in the dog that result in its behaviour. Since each dog has a unique history, however, you will always be better at explaining the dog's response after it happens than predicting what it will be in advance.

It is one thing to describe an animal, person or a plant as a system. The boundaries and the internal routes of communication and control are clear. We can easily identify what systems people call emergent properties – that is, characteristics of the whole that are more than simply the sum of the parts. More than that, emergent properties can only be described in terms of the whole. An attack by a dog is an attack by something more than a bunch of chemicals; it's a bunch of chemicals organized in particular ways and with a collective history and memory. Not all dogs will bite with equal probability; every good veterinarian knows this as she examines the dog and prepares to inject it. In similar ways, we might talk about people as systems. Any living thing may be thought of, not just as a system, but as a self-organizing system. Because of the way the elements connect and communicate with each other, the system organizes itself in certain ways. We may use experimental and epidemiological studies to put numbers on the connecting arrows between animals, plants and people in a particular social or ecological landscape. From this we may calculate the change in human disease that would result from a particular intervention.

But is the larger collection of relationships itself a self-organizing system? Do eco-social communities continually recreate themselves in recognizable ways (you can see I am beginning to insert notions of time and plot here, often missing from systems theories)? Perhaps, but we must add a few cautionary notes. When we talk about systems, many people immediately think about simple or complicated, engineering-type systems (cars, factories) that have well-defined components and can be completely described using mathematical models. A conventional engineering approach to systems assumes that there is one right way to model these complex relationships, and that researchers are capable of figuring out that right way. For an organism such as a dog, this view of things works pretty well.

While this way of looking at systems provides some useful insights, it also comes with some serious problems. Many sociologists are rightly suspicious of systems sciences; if systems are thought to have a right way to function, that right way almost always reflects the ideological dogmas of a given time or place. A family works best if women and children obey the men in the house, some might say, or if peasants obey landowners, and so on. Democracy, equity, human rights – all those weird Western ideas – mess up well-functioning systems. Conventional systems theories can lead quickly down a slippery slope to ethnic cleansing and fascism. Some livestock owners bring adolescent or adult animals into their herds periodically; in North America these are called open herds. Similarly, livestock owners who only bring in semen, but no live animals, are said to have closed herds. Animals that live in closed herds tend to stay healthier than those who live in herds where there are newcomers constantly arriving. Similarly, there is some evidence that people who live in stable, ethnically pure communities are healthier and suffer from fewer infectious diseases than people who live in mixed communities. You get the picture. Well, you get part of the picture, a particular view in terms of particular outcomes through a moderately sized window for a brief period of time.

So what do we mean by a system? There is a large body of systems literature, which I shall not review here, much of it grounded, at least formally, in Ludwig von Bertalanffy's General Systems Theory. Flood gives a succinct review of several varieties of systems thinking in his book *Rethinking the Fifth Discipline: Learning Within the Unknowable* (1999). *[systems theory]*

Because systems theories in biology have historically been linked to an organismic view of nature, some ecologists argue that ecosystems don't actually exist. What exists is a set of relationships among individuals competing for survival. If we can't even agree on whether biophysical systems exist, we run into even more problems when we talk about linked social and ecological relationships. Is the Kathmandu valley a system? Figure 3.2 expands the boundaries of Figure 3.1 and incorporates things like education and economic activity, which the first model externalized, but where are the boundaries? In general, are such apparently loosely or arbitrarily bounded things such as families, farms, and communities systems? Is there such a thing as an ecosystem? In Checkland's terms, how can we define a transformation or an owner of an ecosystem? Is there not a multiplicity of transformations and owners?

Systems theories grounded in complexity, as well as those that focus on systems of thought and systems of inquiry (i.e. systems theory as a way to organize how we think about irreducible complexity), however, free us from the need to give definitive answers to these questions, and still learn something useful. Using systems methodologies, we can begin to account for multiple owners and the many models used to reflect multiple transformations. What we are faced with is not some mechanical system, but a complex reality, which appears to us over time in a canon of manifestations.

As human beings groping our way through a mysterious and messy universe, what more could we ask for?

If we begin to think about complex realities in terms of systems we notice a couple of puzzling and sometimes distressing things. In the first place, all complex realities can be viewed and interpreted from a variety of non-equivalent perspectives. For instance, a systemic description of the Kathmandu valley from an economic perspective will look quite different to that reflecting an ecological perspective.

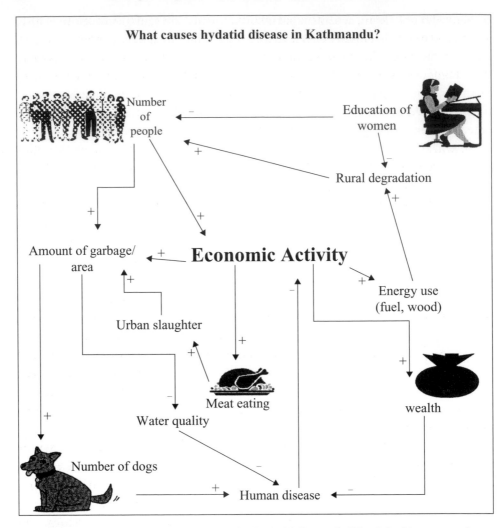

Figure 3.2 Enlarging the boundaries for hydatid disease in Nepal. Is this a system?

The system may function efficiently thermodynamically but be a social disaster, or vice versa. Even within a broad economic perspective, women, children, men and government institutions are likely to identify different elements of importance and draw different systemic diagrams within different sets of boundaries. Furthermore, although they are linked in various important ways, there is no obvious equivalence between these systemic descriptions. There is no right and complete description of the system; there are only various approximations of the mess we live in. In fact, this is one of the key distinguishing features of a complex system versus a simple system, like a car or a computer.

Some professional scholars don't like this, since it seems to admit to a flaky anything-goes view of reality. That is a legitimate, but misplaced concern. I suspect it reflects an underlying personal insecurity on the part of such scholars. Just because the rules of evidence for understanding complex reality are different from those used to understand simple, merely hard reality, does not mean that they don't exist.

I recall hearing, as a child, the story of the four blind men and the elephant.[1] One of them felt a leg and said, 'The elephant is like a tree'. One of them felt an ear and said, 'No, you are wrong, it is like a large leaf.' The third man, wrestling with the trunk, called out, 'No, No you are both wrong, it is very like a snake!' The unlucky blind man at the rear of the elephant sadly moaned. 'Ah, you are so much mistaken. For the elephant is like a rope, and when you pull the rope the heavens open up and cover you with foul dung.' There is an elephant. There are rules of evidence that the four blind men could establish. And one might come up with some collective description – even though we can never be sure that we have the whole, or even the right, picture in our minds. David Krieger, who cites a story with innumerable blind men ('all the blind men' in a Raja's capital city), suggests that the blind men might lead each other around to touch different parts of the elephant and thus enrich their collective understanding (Krieger, 1991). However, even once we have established that the messy elephant we live in has a particular shape and behaves in thus and such a way, there is nothing to say that biology is destiny and even less so that culture is destiny. We can 'change the elephant'. The systems we live in have stories, and while we cannot change the constraints of the stories (birth, death, thermodynamic laws) we can certainly change the plots.

[margin note: We can change the "plots" of our stories]

The reality we live in is considerably more complex than an elephant. Within each perspective, any system can be viewed and understood at a range of spatial and temporal scales. Let us say we are studying farms. We can see that farm households occur within rural communities and watersheds, which are parts of larger ecological and socio-political regions, and so on. Socially, culturally and ecologically, this multi-scalar reality is often best understood in the form of a nested hierarchy, that is, what we have called a holonocracy.

[margin note: holonarchies]

As I discussed in Chapter 1, all living things *can be seen* to exist in holonocracies. Cells are parts of organs, which are parts of bodies, which are parts of families, which are parts of neighbourhoods, which are parts of larger communities, and so on. Each of these is both a whole and a part, and is referred to as a holon. At the small end, it is not always clear that we are dealing with entities that are both whole and a part of something. For decision-making purposes (whether in terms of evolutionary survival or ecological sustainability), it is probably reasonable to

[margin note: holon]

---

[1] There are dozens of versions of this story floating around, This is the one I remember, which may say more about my memory than the story, but that's another story.

begin with organisms (people, animals) as a basic holon. Individual animals and plants have very porous boundaries, drawing from and emptying into their contexts. When dealing with issues of ecosystem sustainability and health, the most useful boundaries are usually spatial. However, some have argued that these are not true boundaries (which may imply barriers), but rather transitions from one set of rules (inside the animal) to another (outside the animal). Still, there is an element of barrier to this transition; for organisms, we recognize the boundaries in part because they are at about the right scale for us to recognize them. At the big end, such as for ecosystems, it is often difficult for people to imagine the boundaries, however porous they may be. Yet, we do behave as if there are different sets of relationships and rules on different sides of transition zones, which, for lack of a better word, I shall call boundaries. Ideas of food webs and invading opportunistic species are based on assumptions that some things belong in certain ecosystems and other things don't. Even if my neighbour and I do not interact socially in any way on a daily basis, our daily lives, how we manage our gardens, how we transport ourselves to and from work, have profound impacts on the spatially defined ecosystem we share.

boundaries [handwritten marginal note]

Early in the twentieth century, physicist Erwin Schrödinger proposed the idea that order arose spontaneously from disorder. More recently, these ideas have been picked up and popularized by authors such as Stuart Kauffman of the Sante Fe Institute. These scholars would argue that membranes around things have been (and are) essential for the development of life. These layers have evolved naturally as ways to keep in check the explosive spontaneous creative reactions that occur when large amounts of energy are poured into chemical mixes (or biological or social mixes). If these membranes were not there, we would have chaos and not life. Thus life exists at the edge of chaos. We tend to think of these membranes purely in terms of organisms. However, anyone who doesn't think communities and cultures have boundaries, which enclose networks of communication and control, must be from another planet. The logical end of globalization, as practiced today, is to draw local, ecosystem-based communities into the sphere of global rules; this is like demanding that the inside of a dog behave according to the same rules as the outside. This happens, of course, when a dog is hit by a car (or skinned alive), which may be an apt analogy for what is happening to ecosystems and communities. You get a lot of exchange with the larger environment for a short period of time, and then the local nutrients and energy are simply absorbed into the larger context.

(a critique of multi-culturalism?) [handwritten marginal note]

One approach taken to studying complex systems is to take one perspective but look across scales. In response to the emergence of food-related diseases such as bovine spongiform encephalopathy (BSE) in the United Kingdom and cyclosporosis in North America, there has been an increasing number of investigations that trace the biological pathways taken by infectious agents from animal reservoirs, through the slaughtering system, and on to the consumer. These studies, called farm-to-fork or stable-to-table within the food safety community, are useful for identifying where in the food chain an organism might be controlled or eliminated. They form the basis for what has been called Hazard Analysis Critical Control Points or HACCP.[2] HACCP analyses are useful under certain limited circumstances that mimic industrial production lines. Usually this means short lines under some single controlling authority – anything from a slaughterhouse to a kitchen. However, these linear methods are weak for dealing with the complex webs of real life. In part, this is because they largely ignore the rule-sets defining holonocratic social and ecological units.

HACCP methods trace a linear path through a complex, multi-layered web of interactions. This works if you have integrated multinational companies who live under the illusion that they can actually control all the variables from farm to fork. Indeed, we have built up whole sets of economic rules based on these kinds of untenable assumptions. Because they rely on control of critical points, HACCP methods applied 'from farm to fork' promote vertical integration of agricultural activities, and they work best when one corporation can control the farmers, processors and distributors. Because the social, ecological and economic webs that feed back into making households, farms or ecosystems viable and self-organizing are not accounted for, our attempts to fit nature into this kind of economic structure are causing massive ecological and social disruption. Farms go bankrupt, streams are polluted and communities appear and disappear. And while agrifood industries that think in terms of HACCP-lines may be surprised by food-borne disease epidemics of salmonellosis, BSE or coliform diarrhoeas, no one with even a passing understanding of complex reality should be at all surprised.

For everything on earth, at least one holonocracy can be imagined; for many things, there are several. Thus, human communities can be seen to be part of nested ecological, economic and socio-cultural hierarchies. As I suggested earlier, these holonocracies may look different if you are a man or a woman, since we will often choose different elements of importance to include in our networks and hence structure our systems view differently. Earlier, I recounted an example from Kenya of how social and ecological systems don't always match up. Those mismatches can be multiplied many times over, depending on the species or communities of

---

[2] Details of how to conduct HACCP analyses can be found on the US Food and Drug Administration website (http://vm.cfsan.fda.gov/~lrd/haccp.html) as well as that of the Canadian Food Inspection Agency (http://www.cfia-acia.agr.ca/english/ppc/haccp/haccp.html).

focus. Surprises occur, however, not just because reality can be seen to be layered, in holonocracies, but because there are multiple holonocracies, which network, or reticulate, with each other. Many of the solutions we devise for public health, environmental, economic and agricultural problems are based on one level and/or one-dimensional thinking; the result of this is that we can create 'surprising' large-scale, long-term problems by solving limited short-term problems in certain ways, or that we can create public health problems by the ways in which we solve agricultural problems. This is both because of cross-level interconnections and same-level interactions and feedback loops, which are inherent in any complex adaptive systems such as those in which we live.

A richer and more realistic approach, therefore, is to incorporate a variety of systems models from different perspectives. How we see a system depends on which elements we choose and how we connect them; the elements we choose as being important reflect what our goals are, and these then constrain the range of options available for policy and management. Researchers, if left on their own, will choose elements of research interest. If we want to influence change in sustainable ways, then we need to create our systems models from those elements that the decision-makers themselves – the stakeholders, the people who live in, and use, the ecosystem – consider to be important. Several ways have been suggested as to how we might do this, some of them using pictures or models, and some using stories. All of them are useful in different ways.

Assessing external relations

Most ways of synthesizing and assessing information (making a diagnosis) focus on the internal functioning of the holon. However, you can learn a great deal about a system by how it relates to the holonocracy of which it is a part. Is the (person, animal, population, community, ecosystem) on an intravenous drip? Or is it interacting across integral but open boundaries? One of the most useful ways yet devised to assess this is by the use of the Ecological Footprint, which was developed by Mathis Wackernagel and William Rees (1996) when Wackernagel was at University of British Columbia. Wackernagel has since taken this idea and made it widely available so that individuals, households, farms or communities can make their own calculations, their own 'self-assessment' if you will.[3] The Ecological Footprint is a way of calculating – for individuals, families, cities, nations – the ecological resources they use.[4] How much of the earth is available per

---

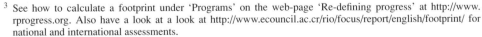

[3] See how to calculate a footprint under 'Programs' on the web-page 'Re-defining progress' at http://www.rprogress.org. Also have a look at a look at http://www.ecouncil.ac.cr/rio/focus/report/english/footprint/ for national and international assessments.

[4] Both the site given above and that at Mountain Equipment Coop (http://www.mec.ca/coop/communit/ meccomm/ ecofoot.htm) have a little calculator.

person? How much are we using? The aim of calculating these footprints is almost always (for Westerners) to reduce them. Going through this kind of exercise can be both humbling and disturbing, especially for anyone who thinks they are living 'green'. By quantifying human impact on our ecological base, the footprint enables us to compare living strategies, set reasonable goals and then work to achieve them.

Understanding feedback loops – from Rich Pictures to influence diagrams

According to Bunch, the Rich Picture drawn by his participants provided a reference point for the workshop and an aid in holistic thinking (Bunch, 2000). Rich Pictures are not only a way for stakeholders to present their situation but also a way to begin making a diagnosis, as different kinds of data are presented in a 'real world' depiction. They can also serve as a basis for designing more formal analytical methods and for proposing further studies to fill the knowledge gaps.

Based on such a picture, we can begin to tease out a more formal set of connections in what have been called 'loop' or 'influence' diagrams. These can be analysed using a variety of formal and informal methods. If you change one particular thing, what are the likely consequences? Which elements have a lot of influences on them? Which elements influence a lot of other things? Which loops lead to accentuating problems? Which dampen them? Richard Levins, an ecologist at Harvard School of Public Health, has described one useful set of methods (see Levins, 1998). Paul Walker, a Principal Research Scientist with the Sustainable Ecosystems Program of the Commonwealth Scientific Investigation Research Organization (CSIRO) in Australia has used the Vensim® simulation package to create simulation models, and examine feedback loops, together with stakeholders.[5]

In my work in Kathmandu, I have found it useful to build loop diagrams in conjunction with community researchers, starting from what they know and/or are concerned about. By doing this, groups of people in a problematic situation can begin to think about 'self-diagnosis'. They can begin to think more carefully about the complexity of their own situation and how they might alter it. We might link open-air animal slaughtering practices, for instance, with the pile-up of offal (solid waste), increased and localized concentrations of dog populations, more dog faeces, and increased transmission of echinococcosis (the stage of hydatid disease which occurs in canines) (Figure 3.3a). We might then layer on links between solid waste and water quality (inverse), and between water quality and gastro-intestinal (GI) disease in general (inverse). We can also add in how education, economic well-being and nutrition relate to these (Figure 3.3b).

---

[5] See http://www.cse.csiro.au/research/Program5/urban_futures.htm

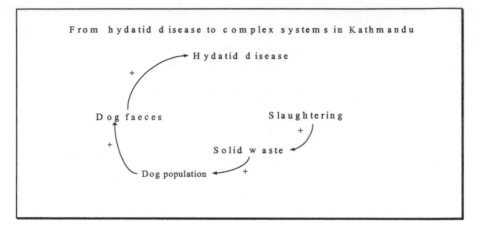

Figure 3.3a Building a sense of systemic interaction in Kathmandu: the hydatid disease path.

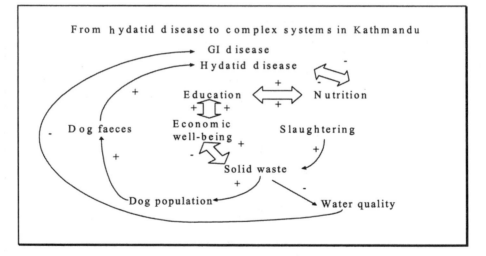

Figure 3.3b The hydatid disease context.

Finally, we can think of some ways – recycling, biogas production, composting and gardening – in which the overall system might be altered to improve water quality and human health (Figure 3.3c).

Or we might ask: how do we decrease GI disease? (Figure 3.4). By giving people better access to taps and toilets. What does that require? Water. Where does the water come from? Rain catchment tanks and groundwater. Can we alter rainfall? Not in the short run. Can we improve our ability to catch the rain? Yes, through better tanks and through planting of trees, shrubs and gardens to prevent rainwater from draining

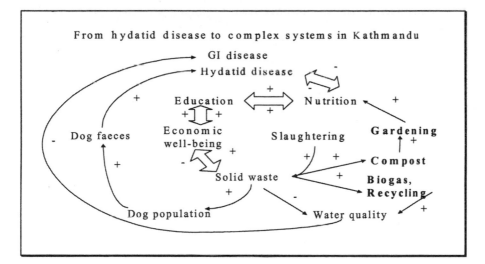

Figure 3.3c Some possible solutions.

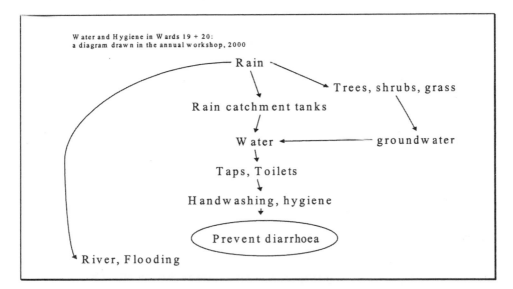

Figure 3.4 Rain catchment tanks and tree-planting to prevent infant diarrhoea in Kathmandu.

directly into the river and creating floods. This greening activity, furthermore, may well help increase future rainfall. By building such models together with the people we are working with, we can gather information and organize it even as we are increasing awareness of interconnectedness and inspiring people to find their own solutions.

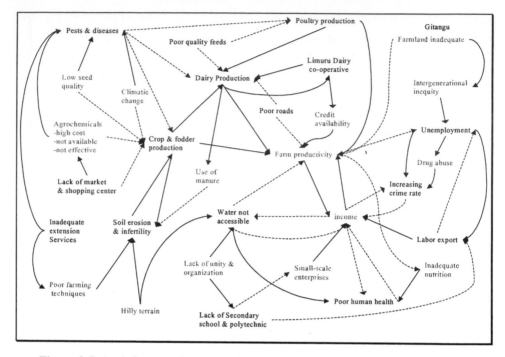

Figure 3.5 An influence diagram of Gitangu village in the Central Highlands of Kenya – original in colour (Gitau *et al.*, 2000).

Thomas Gitau worked with villagers in Kenya in a series of intense workshops and, partly at the workshops themselves and partly by carefully analysing the transcripts of the workshops later, was able to create quite detailed loop diagrams of how villagers saw their ecosystems (Figure 3.5 is an example). They not only identified elements of their surroundings that changed in the same direction or inversely, but also considered whether these changes were desirable or not. Once the diagrams were made, we could ask questions of them, both as researchers and together with the villagers. We can do this qualitatively, to look for strategic points of intervention, and quantitatively or at least formally, to gain greater insight into system dynamics.

We could count how many arrows are going and/or coming from a particular activity or outcome. A lot of links to a goal might indicate more management options. A lot of links from something indicates that it may be very influential. Thomas also assessed the number, direction and degree of feedback loops, with more feedback loops indicating greater complexity. The balance of positive or negative loops in relation to goals indicates stability or instability, with negative loops tending to stabilize a system. Some loops might enhance a goal or objective;

these Thomas referred to as 'regenerative'. If they enhanced a problem, they might be termed 'vicious'. More sophisticated (but not necessarily more realistic) understanding of these diagrams can be obtained if coefficients are available for the various connecting arrows, and you can collect good data over a long period. However, a lot can be learned using relatively rudimentary methods. The villagers of Gitangu drew Figure 3.5. Among other things, they identified links between water accessibility, income and health. The villagers decided to 'turn this around' by focusing on a water development and distribution project, which has been very successful.

Cynthia Neudoerffer has taken this a step further by developing multiple influence diagrams of the situation in Kathmandu, based on both issues (water, food and waste) and perspectives (street sweepers, butchers, vendours; Figure 3.6 shows the butchers' view). We then linked these models in various ways. One of these linked models looked at the relationships between perceptions among stakeholders (Figure 3.7). For example, the street-sweepers complained of over-work, caused in part by people throwing their garbage into the street when they heard the sweepers' bell, but *after* the pick-up tractor had gone by; the community leaders and vendors, however, simply saw garbage in the streets as a reflection of the inefficient work habits of the street-sweepers. Presented back to the group, this linked model offered an opportunity for negotiating a resolution. Another model linked the various stakeholders, their expressed needs, and the resource states identified as indicators of ecosystem health (Figure 3.8).

The loop diagrams suggested so far are all within a particular scale, and tend to work inwards towards greater detail. Some models are more useful for thinking about cross-scale interactions. James Kay, Michelle Boyle and others developed a series of models (Boyle, 1998) that show how the structures and functions of social systems are nested in ecological systems but nevertheless affect their own context and hence change the systems in which they are embedded. (Figure 3.9 shows this at one scale.)

Figure 3.10 applies this model to looking at dynamics among Latin American and North American agricultural policies, diseases and climate. North American demand for cheap food has changed agricultural practices and local climate in southern Honduras in such a way that malaria disappeared, but water and jobs also disappeared and thousands of people headed north looking for work. In the north, heavy spraying by plantation owners to control pests on bananas and pineapples inadvertently selected for resistant malarial mosquitoes. Honduras went from 20,000 reported cases of malaria in 1987 to 90,000 in 1993, most of them in the north (Waltner-Toews, 1999; Almendares *et al.*, 1993). Barrett (1995) has described similar kinds of dynamics for Guatemala.

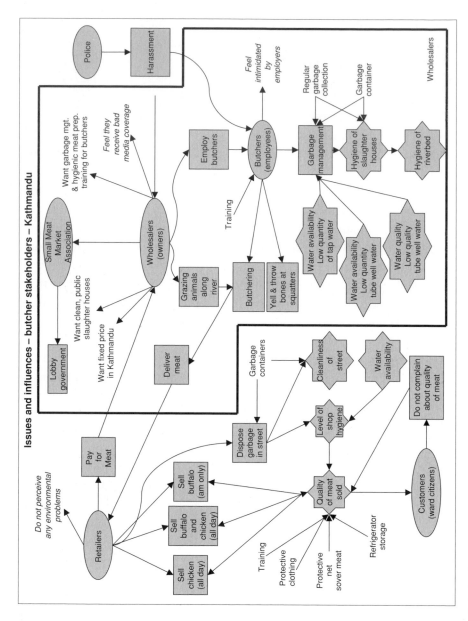

Figure 3.6 An influence diagram of issues and influences in Wards 19 & 20: the butchers' perspective (Neudoerffer *et al.*, 2001).

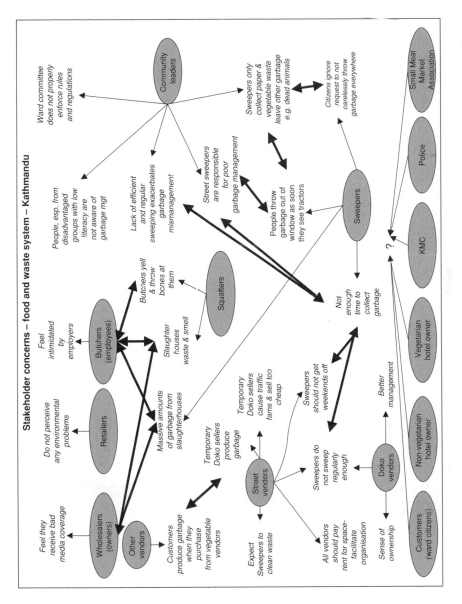

Figure 3.7 An influence diagram of stakeholder concerns in Wards 19 & 20: food and waste system (Neudoerffer *et al.*, 2001).

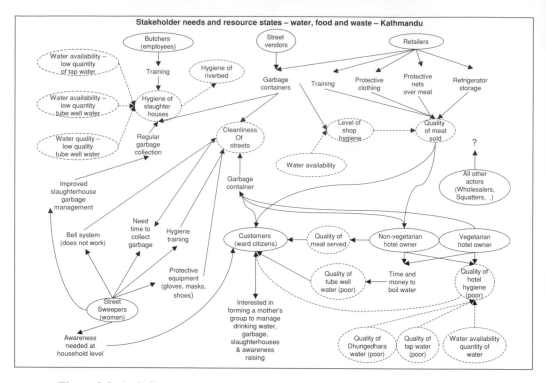

Figure 3.8 An influence diagram of stakeholder needs and resource states (ecosystem health outcome measures) in Wards 19 & 20 (Neudoerffer *et al.*, 2001).

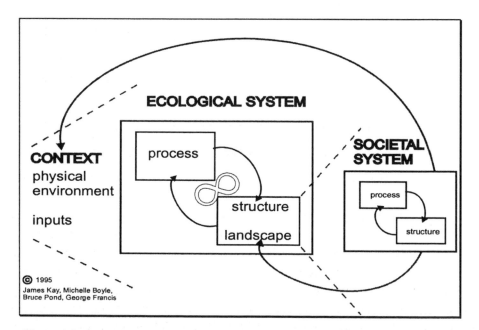

Figure 3.9 A conceptual model for social and ecological systems interactions (one level) (Boyle, 1998).

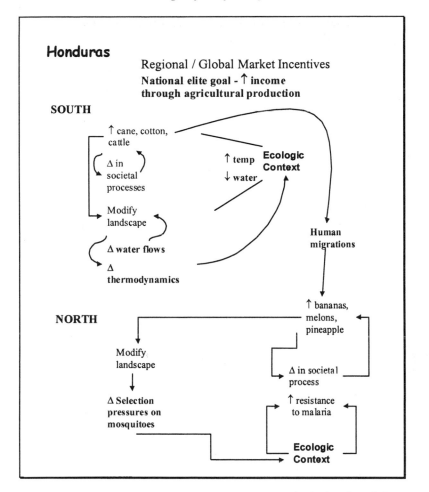

Figure 3.10 How regional market forces can cause ecological and social system changes, and result in the emergence of malaria epidemics.

Understanding self-reinforcing behaviours

C. S. Holling's Lazy-8 and Ulanowicz's partial G clef

C. S. ('Buzz') Holling has put forward a lazy-8 depiction of how many ecosystems seem to work, which has been adopted and adapted by various ecosystems researchers (Figure 3.11). Ecosystems, according to this model, go through phases of exploitation of resources and growth (what in conventional ecological terms is succession) to a mature stage where resources are conserved, through creative destruction (usually a localized collapse), to renewal and reorganization, and finally back to exploitation and regeneration. Holling described the build-up in the ecosystem in terms of stored 'capital'; Kay and others have described it in terms of stored exergy, or useful energy (Kay *et al.*, 1999). This model has been used to describe how forests respond to fires and spruce budworms, as well as how some aquatic

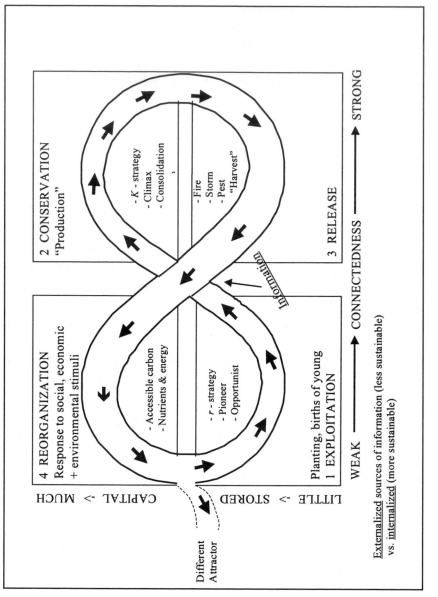

Figure 3.11 Holling's Lazy-8: a cyclic representation of ecosystem development (Holling, 1986).

systems respond to changes in pollution levels. For instance, spruce budworms destroy a small part of the forest, but the genetic information for renewing that area remains local, so the various species can regroup and renew. If we in our wisdom control the budworm over very large areas, then, when an outbreak hits (and it always will) we are faced with a major disaster. The most serious aspect of this disaster is that, since we aren't talking about small, destroyed patches surrounded by large tracts of viable plants and animals, the capacity for renewal has been lost. Holling has suggested analogies in a range of fields, from psychology to economics.

Holling's model is similar to the kinds of mental models used by farmers. They enter certain kinds of information (desirable seeds and genetic information) and suppress other kinds (pests and weeds); they harvest the produce that accumulates during what in a natural system would be called succession; and then reorganize their farm activities around new sets of information. Champions of industrialized agriculture, which is designed for economies of scale, focus only on the development from stage 1 to stage 2 (from Exploitation to Conservation, or planting to production). They do not understand that, by enlarging the space in which the necessary 'release' takes place, and by decreasing the diversity available for response, they are creating the conditions for large-scale plant and animal epidemics, as well as international epidemics of food-borne diseases. Thus the large-scale devastating epidemics of pig diseases in the Netherlands and Taiwan, epidemics of salmonellosis or *E. coli* in food or fungal infections in crops, are, as scientists are wont to say, 'not unexpected'. The way to avoid such disasters is to keep around as much diverse information as possible so that when the Lazy-8 hits stage 1 (Exploitation) and moves into development/succession/increased connectedness, the information is available to rebuild and renew the kind of system we want. This is what is called self-organization. Take away the diversity and you take away the very basis of ecological integrity and renewal. Put in different kinds of diversity (invasive species for instance, or new genetically modified crops) and you may push the whole system into the domain of a different attractor – a whole different Lazy-8 cycle.

Indeed, we need more than what has been called 'requisite diversity', because we need enough diversity (information) to be able to adapt to situations which have not yet occurred. The problem with efficiency is that it tends to push us all towards minimal diversity and hence maximizes probability of failure in the face of change. The principles are the same whether we are talking about food safety, public health, agricultural production, spruce budworms or forest fires. This way of organizing our thinking about the complexity of the real world pinpoints where and why we need genetic biodiversity (to keep open the options for future development), economic diversity (to keep open market options, etc.) and social diversity (to keep open a range of ideas/options/ways of doing things) that will allow the farm or community to adapt to changing times.

Holling's figure also has deep resonances of creation and destruction in old mythologies and religions (which surely themselves reflect intuitive understandings of nature, deeply rooted in our genetic make-up and evolutionary history). It is this that has made it a useful heuristic for drawing lay people and scholars together to identify what particular 'complaints' such as fires and disease outbreaks might mean in the context of normal eco-social systemic development, and to explore possible prognoses and appropriate courses of response.

I have discussed Holling's Lazy-8 at some length because it is widely distributed and used. An intriguing variation of Holling's model was developed by Robert Ulanowicz, which he based on the development of self-reinforcing patterns seen in aquatic systems. The model is topologically equivalent to Holling's but 'resembles a forward-leaning and incompletely drawn G clef from musical notation' (Ulanowicz, 1997: 91). He includes the same four points of reference (renewal, exploitation, conservation and destruction) but his axes are labelled differently. The vertical axis represents the total amount of biomass in an ecosystem. The horizontal axis represents 'mutual information of flow structure'. This is a calculated variable, which reflects the constraints exerted on energy or nutrients passing from one 'compartment' to the next in an ecosystem. Ulanowicz has explained this idea, and a related one which he calls 'ascendency'. He has also demonstrated their theoretical and practical usefulness.

In brief, biomass increases initially very rapidly, and independently of connections (one might say, for instance, lots of growth but few species interactions); as the abiotic resources are used up, the system begins to exploit its own internal resources, and many more connections occur as outputs from some species provide inputs for others. As the system matures, pruning favours fewer but 'more efficient and conservative' trophic pathways, that is, there are fewer but more efficient and more active exchanges between a select number of 'compartments'. It contains a lot of 'mutual information flow structure' in that the interactions are not random, but are constrained by structured pathways, and hence full of 'information'. In its final stages, the system has developed a somewhat rigid but efficient structure, with little room for creativity (the biomass all being tied up in specific structures) and hence is 'brittle'. This is where Holling would say that 'creative destruction' enters. Ulanowicz, in his consideration of aquatic systems, is not convinced that this is inevitable. It may be that ecosystems can make repeated small adjustments. In my view, those 'small adjustments' may represent Holling's Lazy-8 at a smaller spatial scale, within the larger system. They may represent a more conservative and hopeful view of the possibilities for a transition to sustainability.

By considering components of his model called 'ascendency' (organized complexity, calculated by multiplying total system throughput by the constraints on

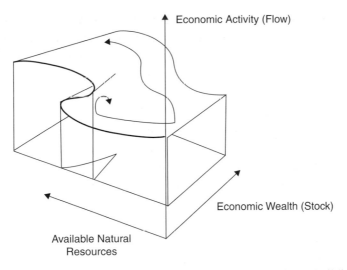

Figure 3.12 Perverse resilience in Peru. Paths to sustainability (building local stock) and unsustainability (exporting local stock without replacement).

that throughput) and overhead ('disordered' complexity), Ulanowicz has used his model to define what 'ecosystem integrity' and 'ecosystem health' might be.

The point is not that one of these models is 'right' and one 'wrong'. They are useful heuristics, and, by structuring how we think about complex eco-social issues, we can learn a great deal about the nature of the complexity around us.

Catastrophe models

In many cases, complex systems will experience sudden, discontinuous changes, which can be described in terms of catastrophe theory (first developed by René Thom). A SOHO system may stay in certain attractors characterized by, say, certain disease rates, species or economic relations. If some variables are changed sufficiently, the whole system may shift to a different attractor in sudden 'flips'.

Figure 3.12, a representation of developments in the department of Ucayali, Peru, based on work by Gilberto Gallopin, James Kay and others, suggests that increased economic activity, drawing on natural resources, may be flipped into a sustainable and healthy state if it is associated with building up local organizational and economic infrastructures. If resources are drawn away from the local system, local resources will be used up and the system will stay close to the x–y axis, and be unable to 'jump' across the fold in the domain space; Ucayali would thus stay in a state of 'perverse resilience' and/or eventually collapse, or reorganize itself around redefined resources. If the forests disappear, perhaps mining would become important, or gravel extraction. This of course has important consequences for species extinction and long-term sustainability. A non-sustainable path, such as the one historically followed in Ucayali, has high economic activity and high natural resource use, but none of the economic wealth stays in the region; it all heads out to Lima and beyond.

A similar model has also been shown to be useful in describing the ecological changes in several of the Great Lakes (see Kay *et al.*, 1999), where phosphate pollution changed the ecosystem from a benthic one to a pelagic one, and pollution control caused a rapid 'back-flip' to a pelagic one. Species were marginalized in both flips, and there were 'winners' and 'losers', since different people had developed livelihoods based on the species distribution in one system state or the other. Thus some sport fishers, unimpressed by the fish species that prospered in the unpolluted system, favoured a return to a more polluted state in order to bring back some of the species they deemed desirable. Of course, since those species are part of a whole constellation of interactions (the attractor) it is not possible to make such gradual shifts. The different species composition patterns exist on different sides of a very sharp dividing line. Increasing pollution would likely not have many impacts until that critical point was reached, at which time more species would be lost as the system reorganized around different trophic pathways.

eg. increasing food-borne diseases

It is also possible that the dramatic increases in food-borne diseases seen in most industrialized countries from the mid-1980s onward are the natural outcomes of an agri-food system that crossed a critical threshold in pursuing economies of scale, integration and international trade. If this is true, it will not be possible simply to reduce the incidence of these diseases gradually, without major reorganizations in the system. Indeed, while intensive efforts and regulations appear to have decreased disease rates slightly, they now appear stable at relatively high levels.

Thus, as with other kinds of models, these catastrophe folds can provide a broad basis on which to explore sustainable or healthy pathways to desirable states.

More sophisticated tools: their uses and limitations – dynamic systems models, spatial models

The loop models and influence diagrams I have described can be formalized into dynamic systems models, with coefficients attached to the connecting arrows. These are useful both analytically and for synthesis. A great many books (see Puccia and Levins, 1985, in particular) and at least two good computer software packages are available that enable advanced students and practitioners in this area to build and assess these kinds of models, and it is not my intention to repeat those here. For the software, and some basic introductions, check out the websites for Stella® (http://www.hps-inc.com/) and Vensim® (http://www.vensim.com/). As I mentioned earlier, the latter has been used extensively by Paul Walker and his colleagues in Australia for qualitative, interactive modelling with stakeholder groups. The more quantitative modelling techniques can be used for examining various subsets of the complex reality we are assessing. Like quantitative patho-physiological and ecological models, however, they are tools for exploring possibilities rather than descriptions of the eco-social systems we are attempting to live sustainably within. Like many systems modelling techniques, they tend to be more useful for studying problems in corporations and businesses than for making diagnostic syntheses in our real, messy world.

Mario Giampietro and his colleagues have developed sophisticated models of what they call 'societal metabolism', in which they integrate insights from complex systems-thinking with biophysical analyses of economic processes. These models and their application are described in detail in two issues of *Population and Environment* (vol. 22, nos. 2 and 3, November 2000 and January 2001). These are beyond the scope of this book, but they are well worth investigating for those who wish to delve deeper into the possibilities of an eco-social physiology.

It is important to build models in as clear and sophisticated manner as possible, *and* to be able to explain them in plain language to the people who want to use them as input for making decisions. Too often, such models remain as esoteric, discipline-based black boxes. Finally, it is important never to confuse the models with reality. Thomas Gitau, whose models I cited earlier, worked with villagers in Kenya to develop some wonderfully complex and accessible models, which the villagers then used for diagnoses and decision-making. The same models could be analysed using more formal scientifically based methods. Similarly, the multiple models which Cynthia Neudoerffer developed were presented to the Kathmandu community meetings for consideration and feedback. They didn't seem to have trouble understanding them.

Other ways of seeing

So far, all the models and pictures of reality I have presented are rooted in a kind of Western materialist view of reality. But Western rationalism is not the only legitimate view of reality. Indeed, some of us would argue that Western rationalist thought, having placed people firmly within an evolutionary and ecological context, must reject pure rationalism as being a preposterous kind of arrogance. We have no way of verifying our view of reality other than by suggesting that it 'works', but of course other ways of looking at the world also 'work'. Some of them, which involve ideas of sacred groves, and ancestors living within our environment (they do, of course, in terms of physical molecules), actually work better for promoting conservation and ecosystem health than biomedical rationalism. Our task in learning our way into a sustainable future is to go beyond postmodern multi-perspectives, however, to find a common, collective narrative for the biosphere.

All cultures are rooted in mythologies, which are expressed in pictures and stories. Many of these provide ways of understanding reality that – in one of those perverse contradictions of logic – are in fact closer to the scientific evidence than the scientific world-view itself. Many anthropologists and ethnographers have delved into this at some depth. One book that explores some aspects of this relevant to our discussions here is *The Spell of the Sensuous* (1996) by David Abram. A philosopher, ecologist and magician, he explores the role that shamans play in traditional societies to mediate human society with their ecological contexts, and the dependence of human cognition on the natural environment.

Margaret Robertson and Pierre Horwitz and their colleagues in Australia have argued that environmental narratives rooted in a sense of place are essential to restore degraded landscapes (Robertson *et al.*, 2000; Horwitz *et al.*, 2001). These narratives, sometimes dismissed by scientists as anecdotal and unreliable, in fact form the essential basis for understanding the 'symptoms' (a sense of vulnerability and loss, for instance), which are different to the biophysical signs measured by scientists. These narratives form the basis for the resilience of individuals and communities – the ability to adapt to change and recover from trauma. They thus lead directly into the discussions of health and management goals that are the subject of the next chapter.

We often think of these place-paced narratives in terms of traditional or aboriginal knowledge. Even in Western scientifically oriented societies, mathematical models may be less effective than poetry and stories for summarizing the complexities of eco-social health. The following is a poem that I wrote and presented at the International Society for Veterinary Epidemiology and Economics; it was also published in *Preventive Veterinary Medicine*, and included in *The Fat Lady Struck Dumb* (Waltner-Toews, 2000b). It represents another way of summarizing a great deal of complex information about BSE.

A bill from the power company
It begins in New Guinea
with veneration of our ancestors.
It begins with love of wisdom and intelligence.
It begins with admiring the dead,
It begins with our cleverness
and our envy.

It begins with eating the brains
of those whom we admire,
with women and children first.
It ends with Kuru, a spongy Jacuzzi
of laid back prions, engulfing the brain.
It ends with a young woman
throwing herself into the fire.

It begins in America with a dust bowl,
with the world war's devastation.
It begins with hungry children
in Europe and Africa.
It begins with our cleverness
and our lust for power and our tractors.
It begins with cows coming in from the green
wilderness in droves.
It begins with praise of hamburgers
on every tongue.
It begins with some spare change
after shopping for food,
and the thrill of a new car.
It begins with recycling, with efficiency,
with cow-eat-cow.
It ends with old men
hungry for power.
It ends with a mob of mad cows
fed to power stations,
shimmering up in smoke
from the incinerators.

It begins with love of life.
It begins in a white coat.
It begins with cutting and sewing.
It begins with new drugs.
It begins with our cleverness
and our fear of death.
It begins with ingesting those
who have what we want.
It begins with blood transfusions,
with hearts, kidneys, and corneal transplants,
with bone marrow and *dura mater*.
It ends with rabies, with AIDs,
with a slow toppling of prions
across the brain.
It ends with a young man,
unable to walk, unable to speak
his own name.

It begins with flowers and wine.
It begins with clever conversation.
It begins with the love of children,
of making them, and, surprisingly, caring for them.
It begins with a house and a car
and a school and a computer.
It ends with paying
the power bill.
It comes round
to huddling by a campfire
playing old guitars,
singing plaintive melodies,
and, in the ash-black darkness at our backs,
the sound of cows or bears foraging,
and the slow sigh of a rising moon.

Triangulation: informing clinical judgement

We talk about 'clinical judgement' in ecosystem health work; however, we need to be careful. Philosopher of science Karl Popper once said that in science there are experts, but there are no authorities. As Funtowicz and Ravetz have shown in their work, the expertise for environmental and health questions is both collective and public; there is no single ecosystem health 'expert', and professionals of all sorts need to learn some humility in incorporating their knowledge and skills into the collective knowledge and skills without trying to commandeer the process. This is not an emergency ward. This is the life of the planet. We all need to be part of the healing process or it just won't work.

The communities and their leaders make the decisions about what they think the most important problems are, and what the best way will be to deal with them. As well, almost all authors in this field acknowledge the need for multiple perspectives. Roe describes this as triangulating – taking orthogonal views of the same situation to arrive at a richer picture of the proverbial elephant. Checkland, in SSM, acknowledges multiple world-views and the multiple models that emerge from those. Cynthia Neudoerffer, James Kay and myself developed multiple models of the situation in Kathmandu. But *then* what does one do? While 'triangulation' seems to provide a way forward, and has the advantage (at least to health practitioners) of mirroring what goes on in making clinical judgements of all sorts – no one really seems to be sure how to 'triangulate'.

How do we integrate the different kinds of information we have gathered? We really are blind people all gathered together, assessing the quality of our information, trying to make sense of the contradictory and complex reality we live in.

Given what we have been saying about the necessity for multiple perspectives, how can we express our collective understanding? At this point, we have no clear rules for doing this, although Raine, Panikkar and others (see Raine, 1998) have proposed various kinds of 'symbolic dialogue', in which the parties attempt to understand each other's world-views. Panikkar's diatopic model for cross-cultural discourse into ecological understanding involves three basic steps:

(1) Within each culture, those who wish to take part in the discourse describe and explain the how, what and why of their traditions and world-views.
(2) Again, within cultures, one traces the origins and contextual development – the ecological, cultural and historical boundaries – of the traditions.
(3) Finally, across cultural boundaries, in the spaces between cultures, we discuss how our world-view is presented in symbolic form, and look for common symbols.

For instance, earth, water, trees and fire have strong mythical and symbolic content in many traditional cultures of the world. They also have strong symbolic content in the cultural beliefs of scientists, where water and energy flows, and notions of physical recycling of nutrients from biological organisms (plants, people, other animals) through the abiotic environment and back into life, form the basis of much of our understanding of ecosystem dynamics. Holling's Lazy-8 model of ecosystem development is so powerful precisely because it taps into deeply rooted notions of birth, death and resurrection, creation and destruction and re-creation. The point here is not to go through a dialectical process to arrive at some universal truth. The point is to work towards a collective, multi-ocular sense of the complexity of the world.

In general, we need to create a space of mutual respect where real listening and real negotiation can take place. This is even more important when we talk about making diagnoses, setting goals and undertaking management of the eco-social communities in which we live. It requires us to be both open and passionate about what we believe is right.

Again, we might ask, 'and then what?' I would suggest that most of us can agree on some general goals: good health and happiness for our children, for instance. We can then allow for a variety of ways to try to attain those goals. I believe that if we have learned to listen with respect, the appropriate paths will be easier to negotiate.

Before we move on, in the next chapter, to consider setting and achieving goals, I think it is worth revisiting the Basic Figure we introduced at the beginning of the book. If we were to redraw it at this point, incorporating the complexity we have introduced, it might look something like Figure 3.13.

I will present another version of AMESH in the final chapter. I present this interim version here to emphasize a few features: we need to work on the scientific

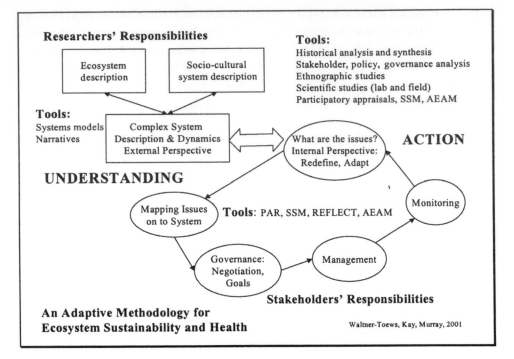

Figure 3.13 A transitional version of the Adaptive Methodology for Ecosystem Sustainability and Health (AMESH), emphasizing roles and methods.

and social sides of the investigation simultaneously; once we have arrived at a workable understanding of the eco-social system, we begin to make a shift from investigative work to management and governance, from research to action. As I shall come back to later, management and monitoring are actually integral to this kind of research. Nevertheless, it is useful to tease apart some of these elements as we go along in order to improve our understanding.

Questions

Undertake several systems descriptions, in terms of connections and feedbacks, scale and perspectives, which incorporate all the information you gathered using the studies in Chapter 2.

Revisit your SSM descriptions and/or your influence diagrams. Try linking them together. Do you see areas of conflict? Commonality? Trade-offs?

Describe your system using a Holling-type Lazy-8 model. Have a look at James Kay's exergy version or Ulanowicz's model and see if that changes your description.

Calculate an Ecological Footprint for some or all of your system, or for yourself.

Discuss the symbols that are important for the cultures engaged in the project you are describing. Are there common symbols or myths that can serve as reference points?

It has been said that all models are wrong, but some are more useful than others. How can you judge which models to use in an ecosystem health setting? What are the scientific and/or sustainable health consequences of using different models? Should some models take precedence over others?

4

Setting goals: where do we want to go?

Setting goals

We are at an important critical point in the process of ecosystem health investigation and management, one that requires a shift of focus away from complaints and problems (disease) to the question of directional change. Having defined the system and its owners, and made some initial diagnoses along with the stakeholders, we need to look ahead. What's the 'point', the 'goal' of the stories we are telling? Where do we want to go? The literature on intervention into systems strongly emphasizes a shift away from thinking about problems to thinking about desirable and feasible systemic changes. So far in this book, we have taken a broadly medical view – starting with complaints, making investigations and diagnoses – albeit 'chastened' by health ideas. Even Soft Systems Methodology refers to a 'problematic situation'. Since this book is about health and ecosystems, I will argue that our ultimate goal is health. I will also argue that the achievement of health requires an entirely different mind-set from that which is useful for solving medical problems.

"good health" defined   Good health is the ability to achieve reasonable, self-defined goals, which I shall elaborate on later in this chapter. If this is so, then we need to set goals in order to measure whether or not we are able to achieve them. But which goals? Whose goals? If you have been reading this book from the beginning, then you will know that the reasonable goals to be pursuing are those of the people and animals and plants that live in a particular place. Many authors have spoken of the need to negotiate goals among stakeholders and provide some useful insights into how this might be done. In the chapter on gathering information (chapter 2), I touched on this. Return to look at propositions 6–9 in Ramirez's framework (Figure 2.3) for guidance in examining conflict situations. In fact, there is a lot of literature – and some new software (see, for instance, www.smartsettle.com) – on how to negotiate outcomes. What is less clear, in a context of sustainability, ecosystem health and the like, is whether it is possible to set the broad constraints, the collective rules to

protect the integrity of the biosphere itself, which make negotiation possible. The ecosystem approach, with its focus on questions of sustainability, seeks to foster collective goals, but these remain largely undefined or ambiguous in the literature. Are any collectively agreed upon goals legitimate? Are there no ground rules?

Without being simple-minded about it, I am convinced that we can describe themes for our goals based on ecological integrity, sustainable livelihoods, and health. Given the focus of this book, I shall discuss health more fully, but scholars and workers in other areas might think how they might describe their goals, and whether or not the same difficult issues arise in their operationalization (see also Waltner-Toews *et al.*, 2004a).

Health is generally accepted by most reasonable people as a non-negotiable, trans-ideological goal. Indeed, I have found that health – rooted as it is in particular histories and cultures, but universally understood in some basic sense – can serve an important role in developing the new cross-cultural symbolic language which Panikkar suggests is necessary to promote convivial and sustainable human life on this planet.

In a sense, health is sustainable development considered in ecosystem terms. Sustainable health of people and other animals assumes a healthy context – a healthy biosphere. Health has been defined and argued about in many other books and scholarly papers. See Waltner-Toews (2000a), Waltner-Toews and Wall (1997) and VanLeeuwen *et al.*, (1999) to pursue this further. In this chapter, I just want to tease out some key issues that we need to understand in the context of ecosystem health management.

Health as a supergoal

Health – physical, social and mental well-being – is rooted in the ability and power to accomplish goals. Philosopher Larry Haworth has called this 'flourishing'. In the world as currently constituted – that is, as a complex eco-social system in which it is meaningless to talk about 'social' or 'ecological' systems as if they had any independent reality – this flourishing has both biophysical and socio-cultural dimensions. For plants and most animals, these goals are biologically determined, within socio-economic and cultural constraints set by people. For people, these goals have mainly become culturally determined. For people, then, health is a socio-cultural construct but clearly within biophysical constraints (our bodies, our ecosystems).

Three statements about health have largely informed public policy in the twentieth century. In its constitution, the World Health Organization defines health as 'a state of complete physical, mental and social well-being and not merely the absence of disease or infirmity'. Two later conferences on health promotion elaborated on this. The Ottawa Charter (WHO, 1986) added that the 'fundamental conditions and resource for health are peace, shelter, education, food, income, a stable ecosystem, sustainable resources, social justice, and equity'. Finally, the Sundsvall Statement (1991) asserted that the 'way forward lies in making the environment – the physical environment, the social and economic environment, and the political environment – supportive to health rather than damaging to it'. These statements, while laudable, are difficult to make workable, and suffer from some intractable problems. The idea of stable ecosystems does not reflect our most current understanding of a change-able nature, for instance, and, as I made clear in Chapter 1, making environments supportive to health is not as simple as it appears, since most environments may support and undermine different aspects of health simultaneously.

health is "context specific & negotiated"

When we get down to the details, health, being goal-driven, is always context specific and negotiated. Apart from this general definition and guideline, which we might think of as a Level 1 health goal, is it then impossible to talk about achieving health in some universal sense? I think not. It is an ideal that informs our practice, but perhaps not much more. There are at least two ways to work through this. One way is to acknowledge the goal as an ultimate ideal, but focus on a process that nurtures sustainable, adaptive goal-setting and goal-achieving. That is the point of most of this book. The other way, which I have already mentioned and which is familiar to most health scientists – and which serves as the basis for most global 'health' programs – is to identify the constraints to health and see if we can remove some of these so that people, animals and ecosystems can flourish from the inside out. It is this latter which I shall dispense with first.

What are the constraints to health?

Focusing on the constraints of health is by far the easiest approach, which is why we do it. What are some of these constraints?

(1) **Limitations of available and useful water, soils, air, energy, complex biological environments.** Maybe we should call this 'vulnerability of integrity'. In this, I would include the physical constraints of holonocracies. A healthy population within some larger constraints requires that either some leave or some die in order to make room for creative newcomers. This is true for individuals in populations, but may also be true for households in communities, and communities in regions. Global ecological integrity (self-organization) puts similar parentheses around all species, including people. Some die even as some are born. We can (and must) negotiate across levels in the holonocracy, so that no individuals or communities create problems for their neighbours, and so that we avoid the problems of cumulative effects, where a lot of good intentions get together to create one big, wide road to hell. This happens if everybody stays healthy by drawing on limited water supplies, or energy, or sends their garbage downstream for someone

else to live with. Having said this however, we are still faced with some ultimate limits, and tragedies that go with them. In order to deal with this inevitable tragedy, I would therefore also add another constraint to health: limitations of cultural rituals to deal with tragedy (see below).

(2) **Disease.** This is the obvious one, the simplest, and the one usually tackled first by 'health' agencies such as the World Health Organization. Yet, as I pointed out in Chapter 1, we may actually create ill-health by the way we attack disease. So maybe it's not so simple after all.

(3) **Powerlessness.** This may be related to a whole variety of things, from gender, age and ethnic origin to land ownership and poverty. Paulo Freire speaks of the oppressed (powerless) as acting according to prescribed behaviours. Traditional education reinforces this. The prescriptions are set by those in power according to *their* goals.

(4) **Poverty.** Poverty per se may constrain goals, but often this is simply a reflection of powerlessness. It is possible for a community to be poor and collectively to achieve a great many things.

(5) **Limitations of cultural rituals, religions, music, poetry.** Communities, which are impoverished in these aspects, will be unable to adapt well to changing contexts, and hence will, by definition, be unhealthy.

(6) **Inability to see things systemically in their full eco-social dimensions.** This does not mean that people need to understand everything, only to have a sense of that integrated reality. People or communities that are 'single vision' are highly vulnerable to change in their environments and thus are unhealthy.

As we might expect, the constraints on health have to do with both human activity systems (culture, social organization, economics), with ecology, and with the linkages between them. This is why, early on, we identified the need for two streams of inquiry, one rooted in each aspect of this problem, and why those two streams need to be brought together in some kind of synthesis. It is also what sets the ecosystem approach apart from both 'pure' social activism and 'pure' environmentalism. A second thing to note is that the various constraints to health are inter-related in the real world, and are related directly to the methods we use to address them: how we do things is as important as what we do.

While the removal of constraints to health – through poverty and disease eradication, liberal democratization – is very seductive, it can, however, be a fatal seduction. The differences between focusing on the problems of constraints and 'solving' them and focusing on the interplay of various issues and how these translate into future opportunities, is the difference between medicine and health. Table 4.1 sets out some of those differences.

Medicine is based on analysis of problems, diagnosis, professional authority, hierarchy and compliance with known protocols and disease prevention. Its aim is to prevent a sick or injured dog or person from dying, or an ecosystem from disintegrating or degenerating. After a car accident, I don't want to negotiate with

Table 4.1 *Differences between medical approaches and health promotion*

	Medicine	Health
Theoretical concern	Analysis	Synthesis
Clinical concern	What is the problem? (specific diagnosis)	What are the issues? (context)
Source of credibility	Professional authority (non-participatory, normal science)	Stakeholder negotiation (participatory, post-normal science)
Social context	Hierarchy	Holonocracy
Main implementation issue	Compliance	Conflict resolution
Goal	Disease prevention or treatment	Flourishing
Systems interpretation	Prevent disintegration	Promote self-organization

the emergency-room doctor. Preventive medicine – based for instance on things like vaccines, drugs, or specific behaviours like condom use – still takes a medical approach. It still focuses on controlling the specific causes of a particular disease. Preventive medicine tends to be based on epidemiological studies.

The promotion of health, on the other hand, requires the synthesis of a variety of perspectives in a holonocratic context, in which negotiation and conflict resolution lead to system flourishing. The idea is to create an organism or system that is less likely to get sick – or if it does get sick, one that will bounce back quickly – because it is well-nourished and resilient. If emergency medicine depends for its success on hierarchy, expertise and control, then conversely, health promotion depends on holonocracy, negotiation and adaptation.

Lyme disease provides an interesting example of how health and medical perspectives differ (Figure 4.1)

Lyme disease was 'discovered' in North America in the mid-1970s (it was already known in Europe), as a result of the perseverance and good record-keeping of some mothers from Old Lyme, Connecticut, who wondered why their village had a cluster of what doctors were calling 'juvenile rheumatoid arthritis'. Eventually their persistence resulted in medical, epidemiological and ecological studies which identified an aetiologic agent (a spirochete, called *Borrelia burgdorferi* after Willy Burgdorfer, who discovered it), a 'new' disease, and a complex web within which the disease appeared. Medical and epidemiological researchers uncovered a wide range of clinical expressions, from rashes and flu-like symptoms to – in untreated cases – chronic heart problems, arthritis, neurological problems, and musculoskeletal pain.

Ecologists have unravelled a natural cycle involving the parasite, ticks (usually some variety of *Ixodes*), field mice, deer and oak trees. Gypsy moth larvae feed

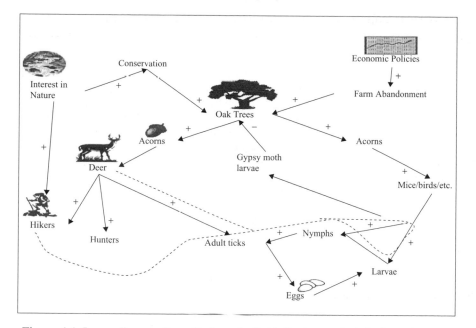

Figure 4.1 Lyme disease. *Borrelia burgdorferi* infection cycle is indicated by the dotted line.

on oak leaves, which is stressful to the trees. Partly in response to this stress, once every three to four years, oak trees put out an extra large acorn crop (called 'masting'). These acorns, rich in fats and proteins, attract field mice and deer that feed on them. The mice also love to feed on pupae of the gypsy moths, which is good for the oak trees.

The ticks follow a life cycle which involves several stages that feed on different warm-blooded animals in this system, and the spirochaete, living in the blood of these animals, is carried from body to body by the ticks. We may begin the cycle with adult ticks, which feed on deer, drop off into the leaf litter and lay eggs which are free of infection. The eggs hatch and the larvae feed on the mice, where they pick up the spirochaete. After a good blood feed, the larval ticks moult into nymphs, which, the following spring, latch onto mice, birds, dogs, hikers and hunters looking for deer – whatever fast food outlet happens to wander into their territory. Late in the summer, the nymphs become adults, and another cycle begins.

A medical approach to Lyme disease involves early diagnosis and treatment with penicillin. Based on a conventional scientific understanding, preventive measures are suggested: wear long-sleeved clothing while hiking, use insect repellants and examine yourself for ticks. Scaled up to the landscape level, one might consider the use of pesticides, depopulation of deer, or major landscape modification. As soon as we begin to consider how one might promote health in this situation, however, we encounter a whole range of contradictory and complex interactions.

A natural history of Lyme disease in North America goes something like this: early settlers arrived in New England, disturbing the natural life cycles of *Borrelia burgdorferi*, mice, trees and deer. The variety of aches, pains and diseases suffered by the settlers were considered 'normal'. As they cleared the forest habitat to make way for farms, the deer populations dwindled away, as did the disease and human memory of it. In the twentieth century, as economic policies took hold that encouraged fewer, larger farms, many New England farms were abandoned. This occurred at a time when the broader (urban) culture was promoting preservation of 'natural' (regrowth) areas (to preserve or promote underlying ecosystem integrity), non-predatory, photogenic species such as white-tailed deer (to help get money for this effort), and outdoor activities such as hiking (to promote both nature awareness and individual health). Furthermore, there is a low medical and cultural tolerance for aches and pains that cannot be attributed to a particular aetiologic agent. These are considered signs of hypochondria.

In a larger eco-systemic health context, then, we are faced with some dilemmas. Promotion of some desirable things (more trees, more wildlife) is linked to the appearance of undesirable outcomes (tick-borne parasitic diseases). Scientists may identify many of the trade-offs, but cannot say which are the 'right' choices. We need to ask what the goals of this system are (including the people) and how we might foster them. We will certainly not abandon preventive medical or treatment approaches. However, new ecosystemic possibilities may suggest themselves, such as increasing diversity of certain species in order to increase the range of animals that the ticks feed on and hence dilute the parasite and lower the infection rates. Such proposals will have implications for many other species – including people – in the system.

Solutions imposed from the outside are usually ineffective at best or, at worst, stimulate a backlash that undermines the original intent. In some situations – though not all – the methods we use to achieve disease control and eradication may undermine our ability to achieve health. The identification of the dangers of tobacco to the biochemical level did not, by itself, lead to a reduction in smoking; indeed, some smokers have resisted anti-smoking laws as being infringements on their freedom to submit themselves to the will of multinational tobacco corporations, regardless of the health consequences. The implementation of centralizing technologies to control food-borne illness has been accompanied by epidemics of food-borne illnesses. Organizations based on medical, veterinary or other 'control and command' ways of thinking, no matter how excellent at providing relief in emergencies (accidents, epidemics) are absolutely the wrong ones to be promoting health.

If we don't focus on disease, poverty – the list of our endless complaints and needs – then where can we turn?

What are the positive attributes of health?

In a kind of perverse contradictoriness that characterizes the complex reality we live in, if we are to deal with the problems such as disease and poverty and powerlessness (and we do want to do that), we may need to change our focus away from disease and poverty entirely. Rather than spending all of our time thinking about problems and constraints, we need to begin thinking about assets and opportunities. Can we turn the constraints on their heads, in a sense turning them into opportunities? If so, how can we do this? Certainly any artist (painter, poet, architect) will tell you that the constraints themselves make creativity possible. Materials, energy, the constraints of language are what we work with to create the future.

There are of course limits to water supplies and solar energy. However, if we look at how various issues are related to each other in the complex web of which we are a part, rather than focusing on single goals (getting rid of a particular disease) we can begin to see some light. It might be possible, for instance, to reduce disease, increase local empowerment and wealth, and even increase water availability, improve soils and conserve energy more effectively through such activities as careful planting of complex plant communities, composting, mulching, solar energy projects and improved infrastructure. The constraints we face – apart from global solar energy input – are not absolute.

Furthermore, it has become quite obvious that setting *single* goals and implementing programs to achieve them without regard to everything else is almost always disastrous. We improve the economy and trash the environment. We improve the environment and create social or economic problems. We reduce disease and destroy cultures. So we not only want to play with the constraints (as any good artist would), but also to seek a kind of balance.

Peter Checkland (1981) suggests that we should think in terms of root definitions for our systems, in which end-goals are hedged by certain value-laden constraints that we might think of as level 2 goals. Checkland suggests that, for each transformation in a system, we should think of effectiveness (is this the 'right' activity to be doing?), efficiency (are we doing it with a minimum of resources) and efficacy (does the way we have chosen to do the transformation actually work?). We might describe an agri-food system, for instance, as a system to transform nutrient-deficient people into nutrient-sufficient people by energy-efficient and ecologically sustainable means. One way to incorporate this into the classification matrix I introduced in Chapter 1 is to create a cube, with scale down one side,

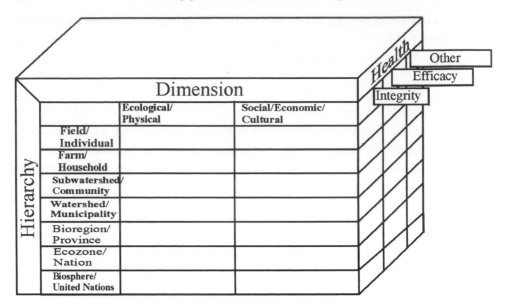

Figure 4.2 A classification cube for research and development by scale, perspective and goals (adapted from VanLeeuwen *et al.*, 1998).

perspective across the top, and health goals in the third dimension (Figure 4.2). Some of us have found this useful for describing the work that we do.

Such goal-descriptors, which can be extended to include ethics, aesthetics, adaptability, and others, are useful, but not a panacea. Some of these, like effectiveness, are problematic in that they require us to determine what is 'right'.

Others, like efficiency, pose other problems. Efficiency shares with effectiveness a common origin in the Latin phrase *ex facere*, to bring about or make happen. By the twentieth century, these two words have diverged. Efficiency gained popularity in the late 1800s and early 1900s when mechanical engineering, economics and managerial science joined forces to create powerful corporate organizations that function according to strict rules of maximizing outputs with minimal inputs. A variation on the efficiency theme was recently introduced with the notion of 'de-materialization', that is, the uncoupling of economic activity from matter-energy through-put (see Robinson and Tinker, 1996). Some, like ecologist William Rees (of Ecological Footprint fame) have argued that this dematerialization is an illusion, that those who increase their income through, say, information services, translate this income into material consumption, which then drives the system that supports them. Many would argue that efficiency has been inappropriately raised to the status of a primary goal in many Western societies, where its role causes social and ecological damage.

Our studies on agro-ecosystems suggest that efficiency does have a place in health assessments but only in terms of its importance for meeting local (e.g. farm-specific) and/or short-term (e.g. seasonal profitability) goals. Managing for efficiency must take place within the constraints of broader ideal goals or the system will become unbalanced and therefore unhealthy.

If we ask whether or not the system is organized to provide a safe and nurturing environment in an efficient manner, we will be required to examine how the organizational structures either facilitate or impede feed-back mechanisms for occurrences of unsafe environmental conditions (e.g. toxins, disease vectors), social conditions (e.g. family violence) or human health conditions (e.g. index cases of infectious disease). For instance, if major decisions affecting the community are made outside that community (as in many agricultural communities, which depend on urban-based industries and governments), then communications regarding such issues as ground-water contamination may have to follow such circuitous routes that the community cannot respond efficiently. Furthermore, if hierarchical (e.g. patriarchal) structures inhibit trust and/or communication, or the community is over-bureaucratized, then social or health problems may not be efficiently reported to those best able to respond to them.

Some of the models set out in the previous chapter – Ulanowicz's or Holling's models for instance – provide useful guidelines as to how one might measure or calculate some of these parameters of health in terms of diversity, dynamics and structure. James Kay and Michelle Boyle have explored this in terms of measuring self-organizational capacities (see www.jameskay.ca for more on this). More recently, members of the Resilience Alliance have set out what they think we should measure; much of their work appears in a series of excellent books (Berkes and Folke, 1998; Berkes *et al.*, 2003; Gunderson *et al.*, 1995) and in the journal *Ecology and Society* (www.ecologyandsociety.org).

Beginning with a local health, environmental or economic issue, the broader socio-economic and cultural context quickly emerges as a significant feature along with the possibility that new organizational arrangements may need to be negotiated in order to deal with that issue. Similarly, the question about whether or not resources are being used in a manner that efficiently achieves a nurturing community would lead to identification of, availability of, access to, and control over local social and biophysical resources. For instance, one might consider whether or not economic resources tied up in spraying repeatedly against certain crop pests (using technology controlled by outside private interests) might not better be used to promote greater ecological resilience in agricultural landscapes, perhaps by designing locally appropriate cropping methods that hinder pest diffusion.

Adaptability, which Checkland doesn't mention but which most of us would consider to be important for health, is sometimes thought of in terms of biodiversity, redundancy, or what Ulanowicz calls 'overhead'. This refers to the ability of the system to respond to, and survive in the face of, a variety of stressors. In this sense, it is closest to the health promotion idea of health as a capacity. For individuals, this may require having a rich variety of ideas and skills, enabling one to adapt to changing social and ecological circumstances. For farms or communities, this may require having a rich variety of economic and natural resources available and a flexible set of organizational structures, which can allow access to and use of those resources under a variety of circumstances.

While we may view adaptability for individuals in terms of their ability to maintain their physical and psychological identities, there may be debate about how to interpret deep structural changes in communities. Is the death of agricultural communities merely a scale issue, which can be viewed as similar to the loss of individual animals in a herd, events which may be tragic but which are necessary to sustain regional health? A key difference, of course, is that the individual animals that are removed from herds are replaced, whereas communities – and the agro-ecosystems which are their creation and for which we depend on food – may simply disappear. Furthermore, some of the work on complexity and hierarchy suggests that the maintenance of appropriate spatial boundaries (ecological boundaries around regions and socio-political boundaries around communities as much as membranes around cells) are essential not only to maintain internal communication and control, but as a mechanism which has evolved to facilitate evolution and global sustainability.

Goals like efficiency and adaptability can often be at odds with each other. As indicated above, local efficiency can be tolerated in managed systems (as in efficient farms) – but that for the larger system (i.e. the community), adaptability is more important. This was apparent empirically when the very efficient health care system in Ontario, Canada, barely coped with an epidemic of severe acute respiratory syndrome (SARS). The need for larger scale redundancy to achieve adaptability is apparent in Holling's Lazy-8 figures depicting stages of ecosystem development, from mobilization (of nutrients and energy) to exploitation, to conservation, and finally to creative destruction, when the 'cycle' begins anew. If the creative–destructive cycles are localized (as happens deliberately with farmers' planting and cropping practices at field level, or as happens naturally in forest patches attacked by pests or fire), then there will usually be resources available (new seeds, fertilizer, surrounding forests, social safety nets, alternative crops for food grown in nearby farms) to provide 'start-up' information (genetics, food, finances) to enable the local system to recover.

This has important implications for local, national and regional policies in such areas as health and agriculture. The systemic globalization of structures and patterns of activity in agriculture is, in this view, very dangerous, since one could expect failures to occur over wide spatial and temporal scales. Patterns of food-borne illness seen in the past decade reflect these systemic changes, as do patterns of emerging infectious diseases.

I won't explore other goals at length here. Some, like equity and aesthetics, may be subsumed under what Robinson and Tinker (1996) have called 're-socialization', that is 'increasing human well-being per unit of human activity'. Aesthetics and cultural richness in particular may be important (whatever other functions they serve) as ways of providing a variety of information for exploitation during times of radical change, and hence maintaining the adaptability of the eco-social complexity which makes human life possible. It may seem odd to say so, but global agricultural sustainability may depend more on better poetry than on more efficient tractors.

In general then, these level 2 goals are useful but not 'the answer'. Systems may be resilient but inequitable. They may be very efficacious at accomplishing bad or maladaptive things. Or we may set out to accomplish seemingly positive goals (such as increasing agricultural production or promoting food safety) in such a way that we create more problems than we solve (fostering inequity in distribution or anti-democratic centralization of economic control, for instance). Some of this may be avoided if we say that a system must achieve all of these things, or that it is not working 'right', and indeed, that kind of striving after heaven on earth may be exactly what we need.

If we put these ideas of health back into our basic 'health promotion' process, we may need to rephrase some of our diagnostic questions.

As Waltner-Toews and Wall (1997) have elaborated, the diagnostic health question may be phrased in systemic terms as: *are the quality and quantity of internal and external resources sufficient, and is their organization appropriate, for the system (person, animal, ecosystem, community) to meet its goals?* Furthermore, if we think in terms of holonocracies, we can add a clarification: *the health of any system cannot be defined in such a way that the survival of the holonocracy within which it is nested is compromised.* In other words, no individual, household or community can or should set goals that destroy its context. This is what laws are for. So, we don't allow individuals to murder, households to throw their garbage into the street, or communities to hog all the water in a valley or harbour diseases that threaten their neighbours. In the same way, we need global laws to prevent countries and multinational companies from destroying what all of us need to have: long, healthy lives. This is both fundamental and common sense – and at least beyond the individual, not usually considered. We seem to be happy to define human health regardless of what it does to our supporting biosphere. Within these fundamental constraints,

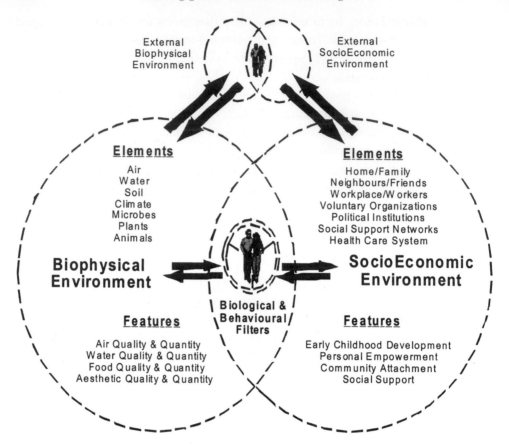

Figure 4.3 A butterfly model of health (VanLeeuwen *et al.*, 1999).

there is a lot of room for specific, achievable objectives. Human flourishing is 'under-determined' – there are lots of ways to become fulfilled as people, and most of them do not involve consumption of the biosphere.

One may also visualize this diagnostic question. John VanLeeuwen and I, along with several others, have developed what we call the 'butterfly' model of health (Figure 4.3). Since we are more interested in promoting sustainable health in a complex eco-social situation, the model draws more on health promotion than on disease prevention.

Some of the features of this model are described in Table 4.2.

The paper by VanLeeuwen *et al.* (1999) gives a detailed description of the origins of this model, and what the different categories and lines mean. Here, I just want to emphasize that the kind of health we seek – the kinds of complex goals we must be setting if we want to get to the root of the presenting complaints we talked about

Table 4.2 *Key characteristics of the butterfly model of health*

Characteristics	Butterfly Model of Health
Socio-economic (SE) environment components	Different human structural elements and functional features influencing, and being influenced by, human health
Biophysical (BP) environment components	Different biophysical structural elements and functional features influencing, and being influenced by, human health
Multiple-species	Multiple biota categories represented
Nested hierarchy	Humans placed inside the BP and SE environment, which are within larger ecosystems, affected by neighbouring and distant BP and SE external environments
Model structure and complex interactions	Arrows and broken lines used to identify, respectively, relationships and permeable boundaries
Self-organization	Positive and negative feedback loops
Location and function of human behaviour and biology	Human population is intimately surrounded by biological and behavioural filters which are affected by, and influence the impact of, the BP and SE environments
Model utility	To describe the health of individual humans, populations, communities and ecosystems
Political influence	Political institutions present as a SE element

earlier – involves a careful balance between social and biophysical aspects of our lives, as well as between local flourishing and global survival. We can certainly, as a species, creatively achieve a great many goals that seemed, only a generation ago, to be impossible. But we cannot be and do everything. We are mortals in a finite world.

Whether we are looking at promoting health, or removing constraints on health, we are talking about some kinds of trade-offs. We may identify many of the trade-offs through influence diagrams. However, assessing these trade-offs in terms of overall health requires another step – evaluating them relative to each other and their impact on the system as a whole. One of the ways that has been proposed to do that is through amoeba diagrams.

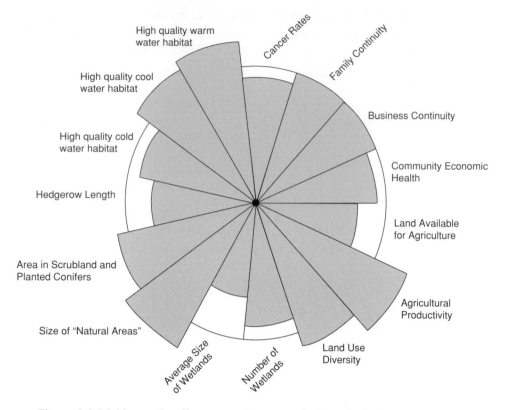

Figure 4.4 Making trade-offs: an amoeba approach (hypothetical agro-ecosystem example adapted from Smit *et al.*, 1998).

Understanding trade-offs – amoeba diagrams

There are various ways to begin to grapple with the kinds of trade-offs we might be faced with in any diagnostic situation. The influence diagram models of Paul Walker in Australia, which I mentioned earlier, are useful in a wide variety of contexts. Another approach, developed and used primarily in Europe, has been to group characteristics into categories we think are important (socio-economic, environmental, self-assessment of quality of life) and then to create 'amoeba diagrams'. In these diagrams, various key indicators are standardized and scaled so that we can see whether the system we are looking at is high or low on the scale, and if this changes when we change other indicators.

For instance, Figure 4.4 gives a hypothetical example taken from the final report of the Agro-ecosystem Health Project at the University of Guelph (Smit *et al.* 1998). The hypothetical agro-ecosystem illustrated in Figure 4.4 has health assessed relative to an arbitrary base for 15 indicators chosen from the array addressed in the Project and each represented by one segment. The human dimension is presented at the top of the diagram and reflects a decline in health status due to elevated cancer rates. With respect to community health, there has been little change in the two indicators for social capital but the economic aspect of community health is shown as declining. Farming system health has also declined in terms of overall land availability for agriculture. Compensating for this downward trend is an increase in productivity and land use diversity.

A number of biophysical health indicators are featured in the amoeba diagram. The number of wetlands is shown to have decreased slightly while their average size has dramatically reduced. By contrast, the size of 'natural areas' (i.e. those with forest cover) have increased, including areas of scrubland and planted conifers. Changes in the characteristics of hedgerows indicate their length has slightly declined. In this hypothetical case, stream health has improved in warm and cool water habitat but has deteriorated slightly in cold water habitat.

When considered overall, the shaded area summarizes or represents the overall health of the agro-ecosystem in terms of how it has improved or eroded compared with the specified reference points. Had these been the desired thresholds or ideal standards, the best shaded shape would be a circular form either bounded by the solid line or extending beyond it. Because deviations from that ideal shape are immediately obvious with such a diagram, it is a particularly useful tool. Not only does it provide a holistic summary of agro-ecosystem health that is easily comprehensible, but it also allows for comparison of a number of agro-ecosystems and could provide the basis for community education and awareness programs.

We could design similar diagrams in which we look at other indicators such as water availability for hygiene (a short-term health outcome measured, for instance, by tapped water use) and agriculture (a long-term health issue related to nutrition), balanced against water in aquifers and replenishment (ecological issues with long-term health implications).

Having extolled the positive side of these diagrams, I would be remiss not to point out some serious problems with them. In order to create such trade-off diagrams, we need to decide what we want and/or need to trade off, and to select indicators that we think represent those important categories. We also need to scale them to selected cut-points (the circle) which we think represents the line between health and ill-health. As you can see immediately, the utility of these amoeba diagrams is entirely dependent on a series of value judgements, based – we hope – on the best information available. We are right back in the middle of multiple perspectives, post-normal science and diatopical symbolic languages.

Even if we can get past this stage, however, we are stuck with some technical problems. If, for instance, the variables chosen are not of the same type – some are linear, others ordinal or log scale – then it is not clear that any meaningful comparisons can be made. A small shift in one variable may be comparable to a very large shift in another. Furthermore, some variables are subject to thresholds; the system compensates to a point and then suddenly shifts. This happened with acid rain and many lakes appeared 'normal' up to a threshold and then abruptly deteriorated. Of course, any medical person will recognize this as an analogy to homeostasis.

Despite the problems, amoeba diagrams can, in the right hands, be useful tools. Pastore and Giampietro (2000), for example, have used them to assess the sustainability of Chinese villages under various development scenarios.

Agreeing on the goals

Even if we have defined and measured health, we still need to find mechanisms to promote its acceptance as an over-riding goal, and then to achieve it. At community levels, where direct negotiation is workable, this is possible. Internationally, there are serious problems for two reasons: the first is that the most obvious global governance structures that could provide a forum for this – the United Nations – are vulnerable to political manipulation by economically powerful nations. This has been seen in the wrangling over how to address global climate change, where those nations that contribute most to the problem are least willing to contribute to the solutions. Nevertheless, some kinds of global organizations are needed, and the United Nations is the best we have.

The second problem with global governance may seem to contradict what I have just said, but is in fact complementary. Health can only be achieved from the bottom up, in the complexity of local contexts and diversity; health cannot be 'given' to someone. The best organizational response I have seen to this is the networking of local groups that formed to respond to the imminent passing of the Multilateral Agreement on Investment (which threatened to empower corporations with rights and no responsibilities), and which successfully provided some democratic corrective voices at the World Trade Organization meetings in Seattle. These networks thus undercut certain kinds of global governance but create new ones (see the work of Naomi Klein, especially her book *No Logo* (2000), for an excellent discussion of some of this). These two organizational streams, both global in scope, but different in constituency and goals, reflect the contradictory needs for democracy and decision-making. In a SOHO, both are necessary to ensure diversity of creative inputs (the networks, which function like individuals, species and habitats) and to ensure the constraints within which that diversity has meaning (United Nations).

At all scales, one of the biggest obstacles is the compartmentalization of knowledge (in universities) and of practice (in governments), which result in the kinds of destructive feedback loops we talked about earlier. Therefore, we need to work at creating organizational structures that foster adaptive communication across departmental lines (health, agriculture, economics). There are some signs that, in the wake of some major water-borne and food-borne epidemics, some of this may be starting to happen. We need to keep up the pressure to maintain the momentum of creative reorganization.

Having made what I think is a convincing argument for health as our 'supergoal', I am well aware that there are differing opinions on this matter. Some hesitancy to use health as a supergoal stems from a fear that the very ideas of human conviviality become medicalized. This is a legitimate fear, which is why I so sharply emphasized the differences between medicine and health in this chapter.

Many ecologists prefer to use the term 'ecological integrity'. I would argue that this is most appropriate for dealing with that subset of ecosystem relationships that do not involve human agency – plants, animals, abiotic elements – within which health is nested. These most assuredly form a basis for human activities, and represent a core of what we need to protect. Integrity also lends itself well to setting rules and legal sanctions. As a forward-looking goal, however, I believe that health is more encompassing. Many development specialists have come to prefer the term 'sustainable livelihoods' to refer to those aspects of ecosystems which only involve human agency, with the ecological context serving as a source of assets. Health is one of the few human notions that incorporates and transcends both.

[handwritten margin note: "ecological integrity"]

In all of this, we are seeking a kind of balance and harmony, as in the Butterfly Model. This is not some Utopian Dream we are after. Rather, we are using health as a way of thinking and acting that will allow us, in the words of Rene Dubos, to find 'a modus vivendi enabling imperfect [people] to achieve a rewarding and not too painful existence while they cope with an imperfect world' (cited in Last, 1988).

Questions

Having summarized the complaints and described the patient from a variety of perspectives, noting feedback loops and conflicts, describe how you will determine which goals to set to achieve. What are the commonalities and differences between different views of the system? How will different perspectives be reconciled? What are the trade-offs? How will these be dealt with?

Group the goals and issues you have identified into meaningful categories and draw an amoeba diagram. Explain how you think it would change if one of the variables were changed.

It is all very well to speak of harmony and balance, but what if there are deep conflicts between goals, and power imbalances among those who espouse them? Should some goals have precedence over others? Why or why not?

How are the practice of medicine and the promotion of health related? Are there tensions between them?

How do sustainable livelihoods, health and ecological integrity relate to each other? Which of these most strongly influence people's daily activities? Does this vary within and between cultures? Are there tensions among them? Are those tensions resolvable? If so, how might we begin to resolve them? If not, how do we deal with that?

5

Achieving goals: managing and monitoring

We manage to achieve certain goals, and assess our progress towards achieving those goals, by monitoring indicators. Then we adjust our management to account for deficiencies (in the language of cognition, we detect 'difference'; see Capra (1996) and the large body of cybernetics literature which this reflects). For that reason, I have brought the final two steps in the Basic Figure together into one chapter. Another reason for doing this is that, of the various stages in the process, these are the most political, technical and bureaucratic, and the least 'investigative'. In terms of Kay *et al.*'s 'diamond schematic' (Figure 2.1), we are in the bottom rectangle: ongoing adaptive management. In the first two steps of the Basic Figure, people with research expertise take the lead in facilitating changed understanding. Setting goals and plans is a kind of turning point, in this ongoing process, with the final two steps – organizational and political responses to changed understandings of the world – being led by local stakeholders, political and non-political leaders and civic society. Because of this, they are also the least generalizable and most culture-specific steps of the process.

Ecologists sometimes say that we don't manage the environment, we manage people. There's an element of truth in that, but that's far too simple an answer. I also confess I'm also not sure what managing people means; the term conjures up too much of social control for me to be comfortable with it. Henry Regier, a noted Canadian environmental scientist, once suggested that we are not so much managing as looking for ways to adapt to an ever changing world. Indeed, system theorist Robert Flood, in *Rethinking the Fifth Discipline: Learning Within the Unknowable,* sets out three paradoxes that he believes are fundamental to managing any organization in the context of complexity:

> We will not struggle to manage over things – we will manage within the unmanageable.
> We will not battle to organize the totality – we will organize within the unorganizable.
> We will not simply know things – but we will know of the unknowable. (Flood, 1999, p. 192)

If you think that all of this sounds abstract and a bit mystical, you are not alone. Does management in the context of complexity simply reduce to a mantra of zen sayings or New Age platitudes? Does it – like meditation techniques – help us to become better business people, dictators, slaves, teachers, plumbers, peacemakers and ecologists? Do we simply do what we feel like because we perceive it to be 'right' – feed the hungry, bomb the World Trade Center, attack Afghanistan, plant flowers, meditate . . . ? Based on systems sciences, are we required to do anything differently tomorrow morning? The short answer is: no. The requirement to do something different arises not from the methods, but from the (ethically-based) goals, which then inform the methods. The explicit over-riding goal of the approach elaborated in this book is to nurture a sustainable, flourishing planet, which includes healthy human, plant and animal communities. This goal then determines the shape and content of the research and management methods we use. Not everything is desirable, feasible or good. While scholarly research methods and management strategies, taken as a whole, may be viewed as being neutral, the particular methodology we use, and the methods embedded in it, reflect a set of values. The idea of science may be value-free. The practice of science is never value free.

The methodology, then, reflects the values of the users. Like medicine, landscape architecture, and many other applied activities, the methodology that I have set out in this book has components which incorporate how knowledge is acquired, and components which incorporate how that knowledge can best be put to use. Indeed, the approach I have outlined belongs to a set of related methodological processes that integrate research and action in such a way that they are inseparable. Every action is a research hypothesis; every research program is a policy and management statement.

Much of the modern systems literature, coming as it does from a business management perspective, does sound abstract and ethically neutral, as if producing gas ovens for a Nazi death-camp and solar ovens for a Calcutta slum are equally laudable goals, resulting in increased employment and economic well-being.

I believe that if we put the new complex systems management techniques firmly in a context of the complexity of eco-social systems, subject to the over-riding goal of fostering a healthy, flourishing biosphere in which healthy human communities have an essential role – then this is more than just another way to help your business make money. There are things we not only can do, but should be doing. And there are systems that are better off if we judiciously neglect them.

If setting goals is problematic for ecohealth practitioners, then achieving them is even more problematic. An individual may arrive at some decision to exercise more, read more, interact more with friends, avoid logo-branded food and clothes, walk, cycle, and then act on these things. A community may pass laws to protect environmentally sensitive areas such as watersheds, but other acts of ecosystem promotion,

such as outlawing or restricting certain kinds of industries (car-dependent super-
stores, those that use slave labour, those which are environmentally destructive) are
becoming increasingly difficult in a global climate which champions the rights of
corporations over those of people, animals or future generations. The more encompassing the holon – regions, the biosphere – the more difficult both arriving at
agreements and achieving them becomes. At the same time, many of the urgent
ecosystem health questions – climate change, degradation of water sources, epidemics related directly to the structure of agricultural industries – require global
rather than merely local solutions. How can we move ahead in such a multi-layered,
complex system?

Agro-ecosystem health management: what can we learn?

Agro-ecosystems are a prototype for the kind of ecosystems we are dealing with –
urban, rural or even so-called pristine. We are increasingly living in a world where
social and ecological systems are so closely enmeshed that many of these qualifiers
on ecosystems don't make much sense any more. Some of the methodologies that
were developed for learning our way into sustainable, healthy agro-ecosystems,
then, can be seen to have more general applications. The world is not a wilderness anymore, however much we might wish it. As Michael Pollen articulated so
eloquently in his book, *Second Nature* (1991), the world has become our garden.

Agricultural activity and the distribution of its products occur within a very complex set of social, cultural, ecological and economic relationships. Whole societies,
including urban industrial societies, are built up around particular ways of growing and distributing food, and, in turn, modify the nature of those activities. Thus,
when we observe agricultural activities in a particular landscape, we are not only
observing the relatively straightforward technical practices related to the planting,
nurturing and harvesting of crops, or the husbanding (wife-ing?) of animals. These
activities, and the food processing and distribution practices associated with them,
also reflect cultural attitudes towards nature, society, and the complex interactions
between the two. Thus agro-ecosystems can be understood as 'emergent complex'
systems, that is, SOHO systems whose complexity eludes comprehensive analysis
and prediction because of the presence of people inside them as driving forces.

Given this complexity, how have we 'managed' agro-ecosystems? Can we learn
anything from this?

In the first place, we have tended to focus our efforts at scales where we think
the important decisions are being made *with regard to the goals we have set* –
the farm. The goals we set were efficient production of food, regardless of the
externalized costs. Traditionally, by working only at the farm level we have been
able to externalize some very large social and environmental costs.

In a management context, goals are expressed in some measurable outcome, and people manage to achieve their goals in relation to that outcome. Thus, an individual might measure success by books published, or income, or number of friends or ability to run a certain number of miles. Each of these would have different implications for how people manage their bodies and their lives.

Similarly, farmers who measure milk produced per cow will manage their farms to achieve goals in relation to that measurable outcome. Farmers who measure biodiversity, or total biological production, or milk per acre, would manage to achieve different outcomes. Their farms would also look very different depending on which measurable outcome they chose as their important outcome. In a herd health program, a farmer might notice that s/he is losing money (presenting complaint). This may be because of something internal to the farm (e.g. her cows aren't producing enough milk to pay for their keep) or because of variables determined at another level in the holonocracy (the price of milk is low or the cost of imported feed is high). To keep things simple, veterinarians and animal scientists tend to focus on the internal factors. The farmer may set a goal of producing more milk. In order to achieve that goal, s/he will have to have some understanding of how feeding, reproduction, housing, disease rates and other variables relate with milk production, and which of these is likely to be the most influential negative factor in the farm (making a diagnosis). Based on this diagnosis, s/he can set some operative goals with regard to, say, reproductive rates and feed efficiency, which can then be managed; then, both the primary goals (milk production) and the operational goals (improving reproductive and feed efficiency) can be monitored using standard indicators. So, as the circle of the Basic Figure in the Introduction comes back on itself, the goals are embedded in the indicators that are being monitored. This is all relatively straightforward. However, it ignores the fact that a farmer has multiple goals, some of which may conflict with each other in practice.

Holistic Resource Management (HRM) is an approach to farm-scale agro-ecosystem management and an international support group that goes well beyond this conventional single-mindedness. Drawing on experiences with ranchers and range managers in Africa, North America and elsewhere, Alan Savory, the founder of this 'movement', has outlined three interlocking sets of goals which farm families must decide on, take ownership of, and then manage (Savory, 1988; www.holisticmanagement.org). Farm families, he argues, need to set goals with regard to the quality of life they want. They then need to set production goals (which may include aesthetic and cultural considerations), as well as goals related to the management of the landscape/ecosystem. Their landscape goals need to be articulated and implemented in such a way that their production goals can be sustained indefinitely. This is thus both more complex and more 'common sense' than, for instance, the Ontario Environmental Farm Plan Program (http://res2.agr.ca/london/gp/efp/efpmenu.html), which provides technical guidance for farmers to manage their landscapes in an 'environmentally friendly' manner, but fails to put this into an eco-social context. HRM forces farm managers to consider carefully and manage explicitly the complex set of goals which most of us use, unconsciously, to guide what we do. This avoids the fuzzy thinking inherent in much 'modern' profit-based production agriculture and requires that farmers face head-on the conflicts and contradictions among landscape, economic and social goals. My impression in talking to farmers who have used HRM is that this is a significant shift towards socially, economically and ecologically sustainable agriculture.

HRM, while it accommodates multiple goals, fails in one important aspect. Like the Environmental Farm Plans, it only deals with activities inside the farmer's boundaries. In the year 2000, in the small town of Walkerton, Ontario, more than 2,000 people became ill and seven died, after one of several drinking wells feeding into the municipal distribution system was contaminated with *E. coli* and *Campylobacter* (Anon, 2000; Mackay, 2002; O'Connor, 2002; Woo and Vicente, 2003). The farmer whose land drained towards the contaminated well, and whose cattle were identified as the source of the bacteria, had a carefully designed and well-implemented environmental farm plan. He was, however, unaware that the city had dug the well just outside his property boundaries. In retrospect, it is clear that the problematic situation in which this tragic epidemic unfolded was bounded neither by the farm nor by the city. A view of the area as a set of multi-layered SOHO systems (agricultural, social, political, ecological) was required to encompass the situation. Farms can be seen as part of socio-economic, cultural and ecological holonocracies. The way we make this manageable is not by reducing the task to linear tasks (production, soil, water), as in production agriculture, nor by focusing only on the farm. Of course we must focus on critical holons, but always with an eye to the overall holarchic context. In a sense what we are trying to do is to put farm-level HRM style management into its larger, multi-scale context.

One of the ways which some of us have found useful to make this change of focus is to begin at the neighbourhood community and landscape ecology levels, and then work 'up' and 'down' from there. While working at the farm level still allows the luxury of thinking in single-level terms, working at community and landscape levels tends to force multi-level and multi-perspective thinking. An individual farmer might be able to sustain the illusion of a single-owner business manager. An agricultural community, however, cannot ignore the farmers of which it is at least partly comprised, and rarely are rural communities sufficiently powerful to ignore the larger context of which they are a part. It is for this reason that I believe transferring technologies from individual farmers to other individual farmers in other parts of the world, however well intentioned, is doomed to failure. Indeed, we could argue that this way of thinking about global sustainable development has already failed. By removing 'farm' problems from their eco-social context, and focusing on technical problems of production, we have created massive dependencies, inefficiencies, and social and ecological destruction.

To promote sustainable agriculture means to promote sustainable rural communities in sustainable ecosystems. Nutrition is the social conscience of agriculture, and ecology is its biological conscience. To the extent that either is ignored, we create a monster, which strips the skin off rural communities in order to foster more efficient trade, and mines local resources to create the illusion of greater economic profit.

If agro-ecosystem management as farm management is inadequate to the task of an adaptive ecosystem approach, and if we need to consider multiple goals at multiple levels, how can we move forward? What should we monitor to measure our success?

What to monitor?

For any health practitioner, it will never be enough to describe and understand a set of problems, nor to work with people to arrive at desirable and feasible courses of action (hypothesis tests). Rather, we want to know something about the value-laden question: is the patient getting better? Despite all our talk about multiple perspectives and uncertainty, that is also a question we want answered for ecosystem health. It would be disingenuous to suggest that 'anything goes', or that we don't know what better means. Happier people, fewer sick people, a healthier, more diverse world, cleaner air and water, sustainable livelihoods, equity . . . surely only the most misanthropic curmudgeon would quarrel with these.

Having said that, in order to measure whether or not we are getting closer or further away from our goals, we need to get away from mere good feelings and find some variables to measure, which will give us an indication of this. What will we use to create our amoeba diagrams or other trade-off heuristics? What will we use to negotiate health?

The variables we measure to tell us whether or not 'things are getting better' are called indicators; measuring these indicators on a regular basis to assess status is called monitoring (to be differentiated from surveillance, which is more focused on specific diseases or conditions, as in disease surveillance). There is a large body of literature on monitoring and indicators – often confusing and certainly murderously mind-numbing in detail.

Michelle Boyle and colleagues have identified seven essential elements of a monitoring program (Boyle *et al.*, 1996; Boyle 1998). These include:

1. A set of human goals. Since a monitoring program assesses progress toward achieving goals, these need to be clearly articulated. (I would add that we need multiple goals to reflect multiple perspectives.)
2. A conceptual model of the world, which provides a context for relating indicators to each other and the system overall. Boyle uses a model in which social systems are nested in ecological systems, but which also alter those systems through feedback loops.
3. A set of indicators.
4. A methodology for data collection.
5. A methodology for calculating indicators.
6. A process for synthesis, so that an overall picture of the system can be created and evaluated.
7. A methodology for reporting. The judicial inquiry into the Walkerton outbreak cited earlier identified both falsification of records and lack of clear reporting and response protocols as important contributing factors to the occurrence and seriousness of the incident (O'Connor, 2002).

We have covered many of these items in previous chapters. Here, we want to focus mainly on indicators and their uses. There are complicated and mathematical definitions of indicators. Gallopin (1996), for instance defines an indicator as a variable, 'an operational representation of an attribute of a system; in other words, it is our image of an attribute defined in terms of a specific measurement or observational procedure'. For purposes of applied ecosystem health, this translates into 'those things that we look at to determine how we are doing'.

For organisms, we choose indicators that integrate bodily processes – temperature, respiratory rates, heart rates, distribution of blood cell types – and then get more and more precise as we zero in on a diagnosis. In herd health programs, farmers and veterinarians look at production rates and reproduction indices before spending a lot of time and effort trying to measure feed conversion, breeding behaviour or sub-clinical mastitis. For groups and communities, we also need general screening-type indicators, and then more specific ones. We can start with absenteeism from school or work, reproductive rates, disease and mortality rates by age, and long-term population numbers – and then zero in on more detailed measures.

For ecosystems, we are just beginning to find useful measures, but they usually have to do with the soil erosion, water flows, diversity and types of species present, numbers within species, then moving on to derived indicators such as thermodynamic flows, nutrient flows, and the like. Again, we can then move to more specifics.

In the context of the ecosystem approach we are talking about in this book, we need multi-layered indicators, appropriate for different spatial and temporal scales, that can tell us whether the system overall is improving or deteriorating. And, unlike the more simple environmental indicators, we need to incorporate human communities into them. Like the goals to which indicators are related, then, they have scientific, practical and value-laden components.

Criteria for selecting indicators

Indicators are variables, and indicator-measurement falls within what epidemiologists would call 'testing'. Such testing would include laboratory tests, questions in a questionnaire, physical measurements – anything that can tell us about the state of the system we are evaluating.

There are two broad questions to consider when selecting indicators: (1) Are they scientifically sound? and (2) Are they practical to use? In Checkland's terms these would be stated as: are they desirable and are they feasible?

What are we looking for in terms of scientific criteria? Clearly, epidemiological criteria for tests such as sensitivity, specificity, and positive and negative predictive values are important. Even in medical circles, many practitioners don't realize that a test can have a very high sensitivity (that is, it correctly classifies those who have the disease), but very poor predictive value. For instance, suppose we devised a test to classify farms as ecologically healthy or not healthy. Our intent is to identify those farms whose practices threaten fundamental ecological integrity. Suppose that the test was this: if the farm modified the original pristine landscape, it was, by definition, unhealthy. Of course, all farms would be declared unhealthy. Surprisingly, this is a very sensitive test. In fact, it is 100% sensitive, since all unhealthy farms are correctly classified. However, it is not very helpful in targeting unsustainable farming practices, since the 'unhealthy' category also includes millions of farms that are no particular threat to the ecosystem. This test is said to have a high false positive rate. Even really good tests tend to have high false positive rates when the condition they are measuring is rare, which is why, for instance, screening tests for genetic abnormalities in foetuses done on the general population are more often wrong than right. In looking for good indicators, then, we want those that

(1) measure the systemic (emergent) properties in which we are interested at the scales we think are relevant (validity);
(2) vary with those properties (up or down) fairly closely and without too much time delay (i.e. are sensitive and specific);
(3) can be linked across scales and dimensions; and
(4) are reliable (give the same result for the same state of the attribute).

Experience in herd and community health programs has shown us that we need to think also about some very practical questions:

(1) Can they be measured with a minimum of cost in time, money, and laboratory infra-structure?
(2) Can they be measured in a time frame that allows them to be useful for preventing or curing problems? (This depends on which problems we are considering.)
(3) Can they be measured at the scale that we need them? Pictures of landscapes may be of use to long-term policy planners in government, but of much less use to farmers on a day-to-day basis.
(4) Are they understandable to the people who want/need to use them? In this context, we need to ask who owns the indicators. When we are talking about eco-social systems, this information must be public information rather than private or in-house government information.

Indicators thus have both practical and scientific aspects, which need to be weighed against each other. There is no point having a very precise measurement that no one will use, or that does not measure progress towards the goals we want in a systemic way. For instance, precise measures of antibody titres to a specific disease may be of concern to someone interested in that particular disease; however, if we are interested in the systemic health of the eco-community, more general measures such as the rate of children being absent from school because of illness, or the availability of potable water over time in relation to aquifer replenishment, may be much more informative.

Different indicators for different purposes – long and short term, qualitative and quantitative

Given the above criteria, we have found in our work that we need at least two sets of indicators. We want, first of all, a set of indicators that people in the community can use to measure progress. Therefore, they need to be easily measurable and yield useful results that can be translated into decisions at the appropriate scale.

We also want indicators that are of use for researchers. Research indicators take longer, and are often time-, money- and technology-intensive. Since they take longer to measure and cost more, we should probably select those that give us information about long-term trends rather than shorter-term changes. However, we may want to use research measures for variables that change in the short term (say water quality) at least once, in order to link community-based indicators to our best scholarly understanding of the system. Thus, a villager might test water cleanliness by smell, taste or sight; a city engineer might look at turbidity. A scientific test might be microbiological or chemical. The latter is clearly more refined but needs a lab and equipment. A farmer might run soil through her hands to test for soil quality; a laboratory would do a chemical analysis.

In designing the Caribbean Animal and Plant Health Information Network in the late 1980s, we struggled with the fact that local farmers, often with limited literacy skills, would be collecting information about their very small herds and flocks and orchards, but that this information was also to be used at national and international levels. We devised some very basic record-keeping systems to accommodate this (Waltner-Toews and Bernardo, 1993). In Honduras, a decade later, Erin Sifton and I devised an even simpler system for illiterate farmers. In this case, farmers would put different coloured stars on a large calendar. Each colour represented one of a limited number of different outcomes: sickness in children, death in chickens. Every month, high school students, as part of a class project, would visit the farmers, collect the records, discuss any stars on the calendar to clarify what had happened, and hand out new calendars. This is clearly far removed from the latest computerized records of somatic cell counts which North American dairy farmers use, but it *is* a reasonable first picture of what was happening on these farms.

Even some villager-derived indicators can be problematic, however. In Kenya, villagers didn't want to measure or discuss some indicators that they knew were important (say, mortality rates) for social and cultural reasons. This is not unusual. In Canada's north, aboriginal people may not report certain illnesses for fear of being lectured to by Western-trained medical personnel. When I was working in Indonesia in the 1980s, district governments would sometimes not report water buffalo mortality because this would reflect badly on them as bureaucrats, and perhaps cost them their jobs. When I returned home, one of the public health laboratory managers confided to me that this also happens in Canada with regard to human diseases. In such cases, it becomes clear that local information needs to be complemented by other sources. This is particularly true for ecosystem health measures that may have implications for regulations and restrictions placed on activities by economically powerful members of the community.

Dealing with this requires a commitment to defining and measuring outcomes at various scales; it is not enough to have, say, herd-level indicators, as in a herd health program, farm-level indicators, as in a farm environmental plan, or national indicators, as one might use in a national environmental monitoring program. Indicators at different scales answer different questions, and in a true ecosystem approach, the indicators used must always be nested and contextual.

In Kenya, Thomas Gitau found that researcher-derived indicators, while scientifically sound, were of little practical use at the farm and village level because they were so costly and time-consuming to measure. While they might be more useful at regional and national scales, no one seemed willing to make the financial and logistical investment at that scale.

In the Great Lakes Basin (GLB), some authorities have used modified versions of Karr's Index of Biotic Integrity and the Hilsenhoff Index of Benthic Invertebrates to assess stream health. However, as Dominique Charron (2001) discovered in the course of a major research project to link these measures to agricultural and economic indicators, there was little overall political commitment in the GLB to measure these indicators in an ongoing, consistent way. They were often being measured because particular individuals in particular offices thought they were important, or because individual farmers were trying to deal with trouble-spots. However, local effects, which may be ameliorated by downstream tree planting, are quite different to cumulative watershed effects, which are more difficult to offset and require preventive action.

Provincial, state and national governments sometimes see ecosystem monitoring as simply looking for trouble. If this is a problem in the wealthiest parts of the globe, one can well imagine the difficulties encountered in poverty-stricken areas. Yet without monitoring, how can we learn and adapt?

Who will monitor?

Ideally, the people who need to make the decisions about responding to the indicators should be in charge of the monitoring. Usually they will delegate the actual measurements to other people, but they should ultimately be responsible. In Kenya, we worked with the villagers to set up 'Agro-ecosystem Health Committees', who were in charge of keeping tabs on important indicators. It was clear that much of the incentive for monitoring, however, came from our continued involvement with them.

Policy-makers, decision-makers and managers often do not have the will to monitor. The philosophy seems to be: what the public doesn't know won't hurt them politically. Indeed, uncovering negative changes through monitoring can be hazardous for politicians. As I mentioned above, when I was working in Indonesia in the mid-1980s, some heads of district animal health offices lost their jobs because they reported higher than 'normal' death rates for water buffalo that had been imported – under a special presidential program – from Australia. Of course, the epidemic suddenly seemed to disappear, which made our job as epidemiologists considerably more difficult. The out-of-sight-out-of-mind working principle however, can be equally hazardous. As the Walkerton outbreak referred to earlier illustrated, a couple of thousand sick people and half a dozen deaths can be very potent citizen-mobilizing forces.

We all get tired of course. If we have the money, we can pay someone to monitor and report to us. Money has been drained in various ways from public institutions over the past few decades, however. The water-borne disease outbreak in Walkerton, Ontario occurred at least in part because water-testing activities were privatized, the water manager falsified data, and insufficient attention had been paid to communications between the laboratory and the public health unit. Another option is to create a strong structure and then rely on committed volunteers working together. In some communities, school classes have made monitoring of streams and drinking water part of their curriculum. There is no simple answer to this, but the question must be faced and dealt with explicitly.

Another set of people who should be involved in monitoring are those for whom the indicator is important. Thus women's groups might monitor issues related to their well-being, poor people might monitor equity, and so on. This, of course, would still require support for the disadvantaged so they could participate, and support for analysing the information systemically.

Any program for monitoring we create must address systemic issues. The Walkerton outbreak, which combined heavier than average rainfall with runoff from a cattle farm, poor well construction, a recently down-sized, and fragmented monitoring system, and human deceit and error, illustrated a wide variety of problems that arise when we fail to think systemically. In the context of monitoring, the outbreak highlighted the very serious problem that we face in monitoring for ecosystem or agro-ecosystem health beyond herd or farm scales. In herd health programs, we can narrow down the range of goals, and we have a defined manager who will make the decisions and carry them out. In monitoring the health of the overall eco-social system, we have no such easy mechanisms. We need ways to synthesize or at least bring together a variety of information from a variety of sources to make integrated decisions. Research disciplinarity and government departmentalism make this very difficult. How can we integrate economic, ecological and health data to make meaningful statements about sustainability and health? Only certain politicians (mayors, premiers, prime ministers, presidents) have that kind of integrated responsibility, and they are too distracted with other things (finances, security) and too poorly educated in socio-ecology to be able to facilitate effective decision-making in this regard. They tend to rely on departments to carry out specific mandates in health, environment and development, which can identify departmentally-related symptoms (water quality, disease, economic development), but cannot address any of the systemic forces which have caused those symptoms.

In some ways, the situation is slightly better globally. Responsibility for much of the monitoring has been taken up as a challenge by non-government organizations. Thus, we can put together a kind of global health card from reports of the World Watch Institute (http://www.worldwatch.org/), the World Resources Institute (http://www.wri.org/), and some supra-national organizations such as the United Nations Environment Programme (http://www.unep.org/), the United Nations Development Programme (http://www.undp.org/) and the World Conservation Union (http://www.iucn.org/). Information from these organizations is not well integrated into policy and management at the appropriate scales. Heads of state and even heads of international agencies do not often see themselves as having responsibilities related to the biosphere overall, even though their decisions have huge impacts.

Ultimately, monitoring at all scales will likely devolve to some combination of interested individuals, governments and non-government organizations, with the media playing an important role in asking uncomfortable questions and keeping the public informed where governments fail. A free press and democratically responsive local institutions (as distinct from formal democratic institutions, which may not be locally responsive) are thus fundamental to successfully tackling ecosystem sustainability and health.

Questions

What is the management process, which will be used to implement programs to achieve the agreed-upon goals? Who will be doing what? Over what time frame? What are the lines of responsibility? How will changes and surprises in the situation be handled?

Describe the monitoring program for your ecosystem health program. What will be monitored? Who will do the monitoring? Who will pay for it? What are the reporting mechanisms? How will you assess its sustainability? What are the processes whereby the people in the system can respond to changes in the indicators?

What are the relationships between non-governmental and governmental governance structures in the area where you are working? What are the implications of those relationships for your investigations?

What are the implications of thinking about this as a holonocracy? How can one effectively monitor, and respond to, multiple interacting scales? Are some scales more important than others? Do the scales which appear to be important for monitoring certain variables correspond to particular scales of governance? What can we do if the boundaries of the system to be monitored and for which management decisions need to be made do not correspond to any formal systems of governance?

6

Responding to change: AMESH and the never-ending story

The process of assessing and creating community-based ecosystem health is never ending. Health is not so much a mirage, as Rene Dubos declared, as it is a moving target. Every person and every community and every generation must redefine what they mean by health in the context of an ever-changing ecological context. Our task is to leave as many options open to future generations as possible. This is why cultural and biological diversity are so important. They represent conserving options and the opening of new options – which is why equity and education are so important.

When an outbreak investigation is completed, and the science and the interviews and the digging around in refrigerators and simulation models and laboratory tests are done, it is the task of the investigators to take all the available information and pull it together into a meaningful narrative. In a food-borne outbreak, this story would tell how, when, and by whom the contamination took place, and set it into a larger context. The lettuce was contaminated in the kitchen two hours before being served, and so and so didn't wash his hands because the taps were placed wrongly and management put him under too much time pressure, and the staff lacked education regarding cross-contamination, etc. Similarly, promoting ecosystem health has to do with creating narratives, stories. These stories are not just entertainments. Our lives depend on them.

Ecosystem health narratives are grounded in realities that can be collectively agreed upon. Not all perspectives are equal. In considering alternative viewpoints, many experimentally oriented scientists, who only recognize realities verified by their own methods, might ask: 'Since when did ignorance become a point of view?' The usual outcome of this arrogant stance is to rely entirely on physical indicators relating to water, energy, landscapes, and so on, and to ignore profound issues of social and ecological attachment which have moulded us as a species into who we are, and therefore serve as the basis for who we might become. Management strategies based only on biophysical measures (signs) tend towards the regulatory and even dictatorial. On the other extreme, many common people around the world can retort, 'Since when did Western scientists become the sole arbiters of collective reality?' The usual outcome of this point of view is to say that reality is entirely constructed, and that, since the systems and holonocracies we have talked about are simply tools we use to understand a complex reality, sustainability is entirely negotiated. The world is a kind of inexplicable hologram. In the end, very knowledgeable scholars like Robert Flood conclude 'we really don't know very much about anything and actually never will' (1999, p. 194).

In between, where many of us live, the ecosystem approach struggles to accommodate a variety of views but tries to stay grounded in global solidarity and in something many recognize as common sense. We do have sufficient knowledge about many diseases to either control or eradicate them. We do know that if raw sewage ends up in the drinking water, people who drink that water are likely to get sick. We know how to promote – and hence how to prevent – water depletion and salination, erosion, and even the loss of ozone protection around the earth. It is disingenuous and pompous to suggest that these things are all in our heads. Certainly, we cannot predict the course of the biosphere, but surely we have some reasonable evidence that cutting down all the world's forests and filling our cities with polluting cars is unlikely to keep our options open for future human flourishing on the planet. And we also know that the kinds of eco-social arrangements which make adaptable, viable communities capable of dealing with the unexpected, complex future are not fully pre-determined; that we need to acknowledge vast areas of uncertainty and irreducible dilemmas which are never likely to be resolved; and that, within broad ecological constraints, will need to negotiate and learn our way collectively.

If we look over the stages of ecosystem health management we set out in Chapter 1, and think of them in terms of what we have discussed in this book, we can create a more complex process. AMESH, the Adaptive Methodology for Ecosystem Sustainability and Health, is an approach to promoting sustainable ecosystems and communities, which grew out of several projects in Peru, Kenya and Canada. The basic process has been succinctly summarized by Waltner-Toews *et al.* (2004b). Figure 6.1 gives one version of this process, specifically adapted from the Basic Figure set out at the beginning of this book, informed by the subsequent journey through complexity, and focusing on its role in elaborating sustainable, healthy narratives.

The basic process can be restructured into five major areas of inquiry and action: an entry point to the problematic situation (presenting issues); from this emerges an analysis of the interactions among stakeholders, issues and governance within the (spatially bounded) system we are investigating; this leads to a gathering of multiple stories and pictures from participants in the situation; from this follows

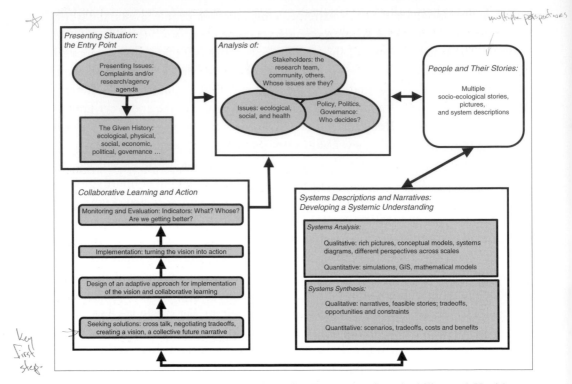

Figure 6.1 An Adaptive Methodology for Ecosystem Sustainability and Health (Waltner-Toews *et al.*, 2004b).

the development of a systemic understanding; and, finally, we arrive at collaborative action and learning (or learning from action). Some parts of the process (gathering of stories, collaborative action) require much closer interaction with the community than others (quantitative systems analysis, for instance). However, in all cases, this must be a truly collaborative effort between researchers and the communities with whom they work.

In the end everyone – researchers and stakeholders alike – should have a deeper, richer understanding of the richness and complexity of the system being examined, and some inkling as to possible ways to learn our way into a more congenial future.

Presenting situation: the entry point

Presenting issues

In step one of this process, we are presented with a complex set of issues, not all of which need be problems. These presenting issues – be they related to water contamination, education, gender equity or deforestation – are embedded in what

Peter Checkland has called a 'problematic situation'. This is our entry point into the complexity of a real world situation, the reason for interacting with some communities rather than others. As we begin to explore this situation, we arrive at a current description of the system, being careful to include considerations of scale, and, so far as is possible at this point, perspective. A subset of this initial exploration is a history of what is there now.

The story so far

As we examine the history, using secondary data and storytelling, government and archaeological records, written histories and genetic and linguistic tracers, we should pay careful attention to the interactions among ecological, social, economic and political developments. What have been the dynamics of change? What has stabilized the system? Who have been the agents of change? At what scale? While much of the recent information (i.e. from the past hundred years or so) comes from secondary sources such as government or church records, in many communities, especially those which have no formal historical records, there are older members who can recount the tale of how this place came to be as it was. It is important to be open to all stories at this point, and also to be sceptical of all of them.

What is the context?

Once we have some historical sense of how the system has unfolded, we are faced with three interacting steps: identification of the issues we wish to address (social, ecological, health), who owns those issues and/or is affected by them (the stakeholders), and the policy and governance organizations that frame those issues and enable (or constrain) solutions.

These three steps in this process are so closely connected that it seems unrealistic to separate them in a particular order. In this case, we shall begin with the stakeholders, since they have often brought the initial complaints.

Who is telling the story?

Who are the real actors and stakeholders in this system? Who is making decisions? Who benefits from the system as it is? Who loses? Who is excluded? In what ways are the stakeholders currently engaged with each other? What are the rules/criteria for engagement or disengagement? What are the coalitions and conflicts? We are trying to find out why the story has unfolded the way it has until now. We want to see who has dominated the story, and if there are marginalized stories which can serve as correctives and provide a kind of insurrection of lost knowledge. Evolutionary changes might be accidental, but the context in which those changes acquire meaning are usually the result of some collectively reinforcing patterns. Again, a variety of methods, from key-informant interviews to workshops, historical records and even newspaper articles can be of use in determining who the stakeholders are. We also need to deal with the researchers and interveners as stakeholders: why are they (we) here? What (or who) gives them power? What is their agenda?

What are the issues?

The stakeholder analysis (who is telling the story) cannot be disentangled easily from the issues that are seen to be important. It is as we work with communities that we come up with a list of the social and ecological issues they think are important, which ones they think are problems, how they are currently dealing with them, and how they would like to deal with them. At this point we are not trying to create complex systems diagrams or even draw feedback loops. We are trying to get all the stakeholders, at different scales and perspectives, to talk about the elements that will eventually go into the systems diagrams. At a local level, we may do interviews, workshops or focus groups. At regional and national levels we may examine various government reports and talk to bureaucrats and politicians. We are trying to work with people inside the ecosystem to come up with a full, rich picture of what is going on.

Policy and governance

Finally, in this interacting mess, we must address policy and governance issues. Here we are going beyond the stakeholders and their issues to look at the rules and policies – both formal and informal – which facilitate and constrain different kinds of interactions. At what scale are the policies and rules formulated? Whom do they favour or hinder? Standard investigative methods for political and policy studies are appropriate here.

Multiple system stories: freeing the narratives

In all the steps so far, and including this one, we are trying to get as many possible versions of reality into the open as possible. I use stories here in a very broad sense, since we may be using maps, drawings, music, influence diagrams, poetry, epidemiological studies, ethnographic studies, and/or economic and mathematical models to present various versions of reality. We would probably use Checkland's SSM to look at the actors, transformations, world-views, owners and environment for each of the various systems perspectives. We also want to define each system perspective by spatial and temporal boundaries, and how it relates to its context.

Thus we may have 'nested' stories. We are not, at this time, trying to synchronize national or regional or local stories. That is a matter for later negotiation. We are merely trying to learn the stories by which people are living.

There are two sides to freeing the narratives, reflecting the two streams of inquiry set out in Chapter 2. On the one hand, we want a sense of all the plants and animals and how they interact on a particular landscape. Are there opportunities to change or use these relationships? Are other attractors possible on this landscape or waterscape?

On the other hand, participatory action processes are essential to each stage of the ecosystem health management, from identification of issues, problems and opportunities, to diagnoses, prognoses and making changes. These processes are essential to free up and give voice to all those who, in the normal course of events, do not have forums in which to make their contributions. Public inquiries after tragedies, such as those in Walkerton, serve this purpose in Canada. However, similar kinds of activities need to be part of any ecosystem health program on an ongoing basis. After the Walkerton tragedy, the provincial government offered the community $5 million to study health issues; then, at the very last minute, perhaps fearing the consequences of a health-focused open inquiry, they withdrew the offer and gave all the money to a medical team studying long-term individual, biophysical consequences of the disease.

The purpose of freeing these narratives is not just to 'give everybody their say'. The purpose of this is to take advantage of the wide variety of knowledge and background there is in any community. We are learning, here, about the biological and cultural diversity, which form the basis of creative future possibilities. It's where we can find new ways to assess situations, and where we can find creative new solutions to old problems. It's where we meet all the characters and think about what kind of a story we might make together. The plot is not yet set in motion.

Systemic understanding

Systems analysis

This 'freeing of narratives' is important, qualitatively, in its own right. It can also be the first stage of an in-depth analysis of various sub-systems from different points of view. With a Rich Picture before us, and a sense of what the issues and problems are, and who the actors and stakeholders are, and how all these things might be linked, we can begin to analyse particular sub-systems. As in Kathmandu, we might look at the water system or the food and solid waste system. One might study energy flows, food webs, predator–prey relationships, community ecology and malarial transmission, sets of economic relationships, or how women interact and work in the community. This analysis may involve detailed studies using both participatory and standard scholarly studies. We are looking at positive and negative feedbacks in each sub-system, and what organizational states are available to them. We are looking for limiting factors (constraints) and opportunities within each subsystem.

Synthesizing whole-system descriptions: constraining the narratives

Once we have a set of pictures, models and stories, we need to link them in some way, without giving dominance to any particular picture or story. One way to do this is through the Rich Pictures, which were introduced earlier. We may also use GIS systems at various scales, with overlays, to see what is happening, and where. We may also use a variety of other formal or informal modelling methods, such as linking initial loop diagrams using Walker's Vensim®-based programs. The idea of using symbolic language, such as that suggested by Panikkar's diatopical methods, would be useful for situations where the cultural differences are great (e.g. indigenous peoples working with urban Western scientists). Basic symbols such as trees and water seem to carry well across some deep cultural divides. We are trying to get a full, rich sense of what is going on in the problematic situation we are exploring. Often, this step and the next two are primarily research activities, constantly informed by, but hidden from, the general public.

Not all stories are possible. We all live within the constraints of a genetic and eco-social past. This is where socio-biologists tend to put all their emphasis. Environmental scientists, on the other hand, often emphasize energy and structural constraints of an eco-social present. Many economists like to point out that the social constructs of economics are major constraints. This is where it is important to keep saying to yourself: Self-Organizing, Holonocratic, Open. The mantra for this work is: SOHO, SOHO, SOHO. Constraints and opportunities are both real and malleable within a nested hierarchical view of nature.

Yes, there are billions of stories that might be told. Within SOHO constraints, perhaps only millions of these are collectively sustainable. Grounded in our best understanding of reality, and our clearest sense of vision as to where we want, or need, to go to achieve sustainable and healthy human communities, we need to identify which collective stories are feasible and desirable.

Here we are asking: Given the range of organizational states available for each sub-system, what are the states available to the system overall? If we undertake certain economic programs, what are the likely ecological consequences? Are the consequences different for various subgroups? Will the women gain something at the men's expense, or vice versa? Will the present generation gain at the expense of future generations?

Action and learning

Finally, because we are dealing with activities of daily life and not just theories, we need to take actions and learn from our actions.

Creating a holonocratic narrative

After all the picturing, posturing, analysis and synthesis, the community we are working with is faced with the big problem: choosing a path into the future. Among the millions of possible stories, we – or rather the communities we are working with – need to choose not just those that are feasible (that is, they work within biological and historical constraints), but are desirable (they take us where, collectively, we think we would like to be). This is the great task of the twenty-first century, in which we all have a stake, for which we need to redefine what it means to be citizens. In the twenty-first century, any concept of citizenship that does not engage us globally and locally as simultaneously ecological and socio-cultural beings is too weak and shallow to be of much use. Building up this sense of citizenship and earth-ownership is one of the great tasks of ecosystem health practitioners.

This step is where our democratic ideals get put to the test as they face the constraints of ecological integrity and political or economic power imbalances. This is where we negotiate trade-offs with as much clarity as we can, look for ways to mitigate expected negative impacts, and create organizations which will enable us to respond if not everything turns out as expected. The kinds of political and social organizational structures that can best accomplish this are only now being invented, and will likely vary from place to place, according to cultural histories.

Ecosystem sustainability and health must be addressed in local, contextual, historical ways, because the species with whom we share this planet have co-evolved in local, historical, contextual ways. But every locality is nested within larger spaces which we need to attend to, and contains within it smaller spaces which we need to nurture. Thinking globally and acting locally is not enough. We must think and act and monitor and react holonocratically.

Making the story happen

Once we have negotiated goals that we want to achieve, we need to create political and social organizations and instruments that can achieve these. Again, these are only now being invented, often in the clash between political leaders and their citizens, between corporate privateers and members of local communities. Both the five-year plan of totalitarian regimes, which assume total control, and the naive 'invisible hand' of anti-planning ideologies have largely been shown to be failures.

We are at a point in human history where we cannot escape our collective respon-sibilities. We need to think hard about where we want to go and create the best instruments to get there. Convivial, sustainable, healthy communities will not occur by accident, nor by total control, but through adaptive, flexible, realistic policies and politics which are not afraid to restrict and not afraid to let go.

Monitoring adapting, re-telling

This process never ends. The universe unfolds and changes, often (usually?) in surprising ways. Therefore, the process of ecosystem health management must become a process whereby we learn from our experience and respond and adapt to the unexpected, reassessing complaints, rethinking the shapes of the ultimate patients, testing different responses. We need good indicators of sustainability and health that not only reflect our best scientific understanding of what is necessary to survive, but also our deepest intuitive sense of what is good and important to survive well.

There is an underlying assumption running through all the activities in this book. Ecosystem approaches to health require that – with all our faults and betrayals and stupidity and brilliance and humility – we need somehow to trust each other. We do not seek a blind trust. Nor do we need to trust each other about everything. We know each other too well for that anyway. We need to trust each other to deal honourably in our public spaces – in the water, the air, the landscape, the way a bird trusts that the branch it lands on will not break, or a fish trusts that water will provide it with oxygen. This is a kind of democratic trust, where we share responsibilities, rights, knowledge and activities, but where we also put into place 'quality control' rules, so we can check up on each other. We have responsibilities not just to work collectively and think systemically, but to be sceptical of everything we do, even as we trust. This is not an easy road, but one, I would think, well worth taking. Even if the world ends – which it will someday in any case – we still want to be able to hold our heads up with dignity as a species, and be able to say we have lived not only long, but well.

Questions

Using all the steps outlined in AMESH, tell the story of a problematic situation with which you have been involved.

What are some of the most difficult or troubling components of this process?

What are the roles of researchers in this process? What are the roles of people involved in governance, whether formal or informal? How do research and development relate to each other? How does this alter the conventional relationships between science and politics? What kinds of problems might this create for both researchers and politicians?

This process is most often described in terms of local communities, yet is theoretically situated in a set of global holonocracies. What special problems arise if you try to apply this process at large geographic scales? Is it possible to do this kind of work globally? Are there global goals, for instance, that we can agree on? Who are 'we' in this case?

What are the assumptions of this process of investigation, for instance, in terms of social organization, political structures, culture, knowledge generation and assessment?

References

Abram, D. (1996). *The Spell of the Sensuous*. New York: Vintage Books.

Allen, T. H. F. and Hoekstra T. (1992). *Toward a Unified Ecology*. New York: Columbia University Press.

Allen, T. H. F., Bandursky B. and King A. (1993). *The Ecosystem Approach: Theory and Ecosystem Integrity*. Report to the Great Lakes Advisory Board, Washington, DC: International Joint Commission.

Almendares, J., Sierra, M., Anderson, P. and Epstein, P. (1993). Critical regions: a profile of Honduras. *The Lancet* 342: 1400–1402.

Angula, F. and Griffin, P. (2000). Changes in antimicrobial resistance in Salmonella enterica serovar Typhymurium. *Emerging Infectious Diseases* 6: 436–437.

Aramini, J. *et al.* (2000). Drinking water quality and health care utilization for gastrointestinal illness in Greater Vancouver. *Canada Communicable Disease Report* 26–24: 211–214.

Barrett, B. (1995). Commentary: plants, pesticides and production in Guatemala: nutrition, health and nontraditional agriculture. *Ecology of Food and Nutrition* 33: 293–309.

Berkes, F. and Folke, C. (eds.) (1998). *Linking Social and Ecological Systems: Management Practices and Social Mechanisms for Building Resilience*. Cambridge: Cambridge University Press.

Berkes, F., Folke, C. and Colding, J. (eds.) (2003). *Navigating Social-Ecological Systems: Building Resilience for Complexity and Change*. Cambridge: Cambridge University Press.

Boyle, M. (1998). An Adaptive Ecosystem Approach to Monitoring. MSc thesis, University of Waterloo. (http://ersserver.uwaterloo.ca/jjkay/grad/mboyle/)

Boyle, M., Kay, J. J. and Pond, B. (1996). *State of the Landscape Reporting: The Development of the Indicators for the Provincial Policy Statement under the Planning Act*. Ontario Ministry of Natural Resources, Ontario.

Bruner, J. (2002). *Making Stories: Law, Literature, Life*. New York: Farrar, Straus and Giroux.

Bunch, M. J. (2000). An Adaptive Ecosystem Approach to Rehabilitation and Management of the Cooum River Environmental System in Chennai, India. Ph.D. thesis, University of Waterloo.

Capra, F. (1996). *The Web of Life: A New Scientific Understanding of Living Systems*. Anchor Books, New York.

Casti, J. L. (1994). *Complexification: Explaining a Paradoxical World Through the Science of Surprise*. New York: HarperCollins.

Charron, D. (2001). Livestock Production and Stream Health in the Great Lakes Basin: an Agroecosystem Health Approach. Ph.D. thesis, University of Guelph.

Checkland, P. (1981). *Systems Thinking, Systems Practice*. Chichester: John Wiley & Sons.

Checkland, P. and Scholes, J. (1990). *Soft Systems Methodology in Action*. Chichester: John Wiley & Sons.

Drotman, P. (1998). Emerging infectious diseases: a brief biographical heritage. *Emerging Infectious Diseases* 4 (3): 372–373.

Flood, R. L. (1999). *Rethinking the Fifth Discipline: Learning Within the Unknowable*. London: Routledge.

Funtowicz, S. O. and Ravetz, J. R. (1993). Science for the post-normal age. *Futures* 25: 739–755.

  (1994). Emergent complex systems. *Futures* 26: 568–582.

Gallopin, G. C. (1996). Environmental and sustainability indicators and the concept of situational indicators. A systems approach. *Environmental Modelling and Assessment* 1: 101–117.

Giampietro, M. (ed.) (2000, 2001). Societal Metabolism. Two special issues of *Population and Environment* 22 (2) and (3).

Gitau, T. *et al.* (2000). Agro-ecosystem health: principles and methods used in high-potential tropical agro-ecosystem. In M. A. Jabbar, D. G. Peden, M. A. M. Saleem and H. Li Pun (eds.). *Agro-ecosystems, Natural Resources Management and Human Health Related Research in East Africa*. Nairobi: International Livestock Research Institute.

Gunderson, L. H., Holling, C. S. and Light, S. (eds.) (1995). *Barriers and Bridges to the Renewal of Ecosystems and Institutions*. New York: Columbia University Press.

Habitat (United Nations Centre for Human Settlements). (1991). *Guide for Managing Change for Urban Managers and Trainers*. Nairobi, Kenya.

Hayles, N. K. (2000). The interplay of narrative and system, or what systems theory can't see. In W. Rasch and C. Wolfe (eds.) *Observing Complexity: Systems Theory and Postmodernity*, Minneapolis: University of Minnesota Press. p. 137–162.

Health Canada (2000). Waterborne outbreak of gastroenteritis associated with contaminated municipal water supply, Walkerton, Ontario, May–June 2000. *Canada Communicable Disease Report* 26-20 (accessed at www.hcsc.gc.ca/pphb-dgspsp/publicat/)

Holling, C. S. (1986). Resilience of ecosystems: local surprise and global challenge. In W. C. Clark and R. E. Munn (eds.), *Sustainable Development of the Biosphere*. Cambridge: Cambridge University Press, pp. 292–317.

  (1995). Sustainability: the cross-scale dimension. In M. Munasinghe and W. Shearer (eds.) *Defining and Measuring Sustainability*, Washington, DC: United Nations University and the World Bank.

Horwitz, P., Lindsay, M. and O'Connor, M. (2001). Biodiversity, endemism, sense of place, and public health: inter-relationships for Australian inland aquatic systems. *Ecosystem Health* 7: 253–265.

Hunter, J. M., Rey, L. and Scott, D. (1982). Man-made lakes and man-made diseases. *Social Science of Medicine*. 16: 1127–1145.

Izak, A. and Swift, M. (1994). On agricultural sustainability and its measurement in small-scale farming in sub-Saharan Africa. *Ecological Economics* 11: 105–125.

Kay, J. J. and Schneider, E. (1994). Embracing complexity, the challenge of the ecosystem approach. *Alternatives* 20 (3): 32–39.

Kay, J., Regier, H., Boyle, M. and Francis, G. (1999). An ecosystem approach to sustainability: addressing the challenge of complexity. *Futures* 31: 721–742.

Klein, N. (2000). *No Logo: Taking Aim at the Brand Bullies*. Toronto: Knopf Canada.

Koestler, A. (1978). *Janus: A Summing Up*. New York: Random House.

Krieger, D. J. (1991). *The New Universalism*. Maryknoll, New York: Orbis Books.

Last, J. (1988). *A Dictionary of Epidemiology*. Oxford: Oxford University Press.

Lederberg, J., Shope, R. and Oaks, S. (1992). *Emerging Infections: Microbial Threats to Health in the United States*. Washington, DC: National Academy Press.

Levins, R. (1998). Qualitative mathematics for understanding, prediction, and intervention in complex ecosystems. *Ecosystem Health*. Winnipeg, Manitoba: Login Brothers Book Company, pp. 179–204.

Linthicum, K. J., Anyamba, A., Tucker, C. J., Kelley, P. W., Myers, M. F. and Peters, C. J. (1999). Climate and satellite indicators to forecast rift valley fever epidemics in Kenya. *Science* 285: 397–400.

Mackay, B. (2002). Walkerton, 2 years later: 'Memory fades very quickly'. *Journal of the Canadian Medical Association* 166 (10): 1326.

McDermott, J., Gitau, T. and Waltner-Toews, D. (2002). *An Integrated Assessment of Agricultural Communities in the Central Highlands of Kenya*. Final Summary Report to the International Development Research Centre, Ottawa (available online from the Network for Ecosystem Sustainability and Health: www.nesh.ca).

McMichael, A. J. (1993). *Planetary Overload: Global Environmental Change and the Health of the Human Species*. Cambridge: Cambridge University Press.

(1999). Prisoners of the proximate: loosening the constraints on epidemiology in an age of change, *American Journal of Epidemiology* 149: 887–897.

Midgely, G. (2000). *Systemic Intervention: Philosophy, Methodology, and Practice*. New York: Kluwer Academic/Plenum.

(2003). *Systems Thinking*. London: Sage.

Minkin, S. F., Rahman, R. and Islam, M. A. (1996). Flood control embankments and epidemic Kala-Azar in Bangladesh, *Ecosystem Health* 2(3): 215–226.

Neudoerffer, C., Waltner-Toews, D. and Kay, J. (2001). *AMESH Analysis of the Urban Ecosystem Health Project, Nepal*. Final Report to the Urban Eco-Health Project (available online from the Network for Ecosystem Sustainability and Health: www.nesh.ca).

O'Connor, D. R. (2002). *Report of the Walkerton Inquiry. Part I. The Events of May 2000 and Related Issues*. Ontario Ministry of the Attorney General, Toronto.

Pastore, G. and Giampietro, M. In M. A. Jabbar, D. G. Peden, M. A. Mohammed Saleem and H. Li Pun (eds.) (2000). *Agro-ecosystems, Natural Resources Management and Human Health Related Research in East Africa*, International Livestock Research Institute, Nairobi, and International Development Research Centre, Ottawa. Also at http://www.nesh.ca/cybrary/ecoh/ultimate-appendix-amoeba.pdf

Pickering, J. (1997). *Health Research for Development*. Ottawa: Canadian University Consortium for Health in Development.

Pollen, M. (1991). *Second Nature*. New York: Delta Trade Paperbacks.

Pretty, J. N., Guijt, I., Thompson J. and Scoones, I. (1995). *Participatory Learning and Action: A Trainer's Guide*. London: International Institute for Environment and Development.

Puccia, C. J. and Levins, R. (1985). *Qualitative Modelling of Complex Systems: An introduction to Loop Analysis and time Averaging*. Harvard University Press, MA.

Rabsch, W., Hargis, B. M., Tsolis, R. M., Kinglsey, R. A., Hinz, K-H., Tschaepe, H. and Saumler, A. (2000). Competitive exclusion of Salmonella enteritidis by Salmonella gallinarum in poultry. *Emerging Infectious Diseases* 6: 443–448.

Raine, P. A. (1998). Sharing ecological wisdom through dialogue across worldview boundaries. *Southern African Journal of Environmental Education* 18: 38–46.

Ramirez, R. (1999). Stakeholder analysis and conflict management. In D. Buckles (ed.), *Cultivating Peace: Conflict and Collaboration in Natural Resource Management.* International Development Research Centre, Ottawa, and the World Bank, Washington, pp. 101–126.

Ravetz, J. (1999). Post-normal science. *Futures* 31 (7), special issue.

Rennie, J. K. and Singh, N. (1996). *Participatory Research for Sustainable Livelihoods.* Winnipeg, Canada: International Institute for Sustainable Development.

Robertson, M., Nichols, P., Horwitz, P., Bradby, K. and MacKintosh, D. (2000). Environmental narratives and the need for multiple perspectives to restore degraded landscapes in Australia. *Ecosystem Health* 6: 119–133.

Robinson, J. and Tinker, J. (1996). *Reconciling Ecological, Economic and Social Imperatives: Towards an Analytic Framework.* Sustainable Development Research Institute Discussion Paper Series 95-1. Vancouver, BC: Sustainable Development Research Institute, University of British Columbia.

Roe, E. (1998). *Taking Complexity Seriously: Policy Analysis, Triangulation and Sustainable Development.* Boston: Kluwer Academic Publishers.

Savory, A. (1988). *Holistic Resource Management.* Washington, DC: Island Press.

Smit, B., Waltner-Toews, D., Rapport, D., Wall, E., Wichert, G., Gwyn, E. and Wandel, J. (1998). *Agroecosystem Health: Analysis and Assessment.* Faculty of Environmental Science, University of Guelph, Guelph, Ontario, Canada.

Sundsvall Conference on Health Promotion (1991). Supportive environments for health: the Sundsvall statement. *Health Promotion International* 6(4): 297–300.

Ulanowicz, R. E. (1997). *Ecology, the Ascendent Perspective.* New York: Columbia University Press, 201 pp.

VanLeeuwen, J., Nielsen, N. O. and Waltner-Toews, D. (1998). Ecosystem health: an essential field for veterinary medicine. *Journal of the American Veterinary Medicine Association* 212: 53–57.

VanLeeuwen, J., Waltner-Toews, D., Abernathy, T. and Smit, B. (1999). Evolving models of human health toward an ecosystem context. *Ecosystem Health* 5: 204–219.

Wackernagel, M. and Rees, W. (1996). *Our Ecological Footprint: Reducing Human Impact on the Earth.* Gabriola Island, BC: New Society Publishers.

Walters, C. (1986). *Adaptive Management of Renewable Resources.* New York: Macmillan Publishing Company.

Waltner-Toews, D. (1999). *Mad Cows and Bad Berries.* Gatekeeper Series 84, London: International Institute for Environment and Development.

  (2000a). The end of medicine, the beginning of health. *Futures* 32: 655–667.

  (2000b). *The Fat Lady Struck Dumb.* London, Ontario: Brick Books, pp. 25–26.

  (2001). An ecosystem approach to health and emerging and tropical diseases. *Cadernos de Saude Publica/Public Health Reports* 17 (1) (Supplement): 7–36 (including critiques and reply).

Waltner-Toews, D. and Bernardo, T. (1993). Record-keeping systems for small-to-medium scale livestock enterprises. In P. W. Daniels, S. Holden, E. Lewin and Sri Dadi (eds.) *Livestock Services for Smallholders: A Critical Evaluation.* Indonesia International Animal Science Research and Development Foundation, Indonesia, pp. 51–53.

Waltner-Toews, D. and Lang, T. (2001) A new conceptual base for food and agricultural policy: the emerging model of links between agriculture, food, health, environment and society. *Global Change and Human Health* 1(2): 116–130.

Waltner-Toews, D. and Wall, E. (1997). Emergent perplexity: in search of post-normal questions for community and agroecosystem health. *Social Science of Medicine* 45(11): 1741–1749.

Waltner-Toews, D., Kay, J. and Lister, N-M. (2004a). *The Ecosystem Approach: Complexity, Uncertainty, and Managing for Sustainability*. New York: Columbia University Press.

Waltner-Toews, D., Kay, J., Tamsyn, M. and Tamsyn, C. (2004b). Adaptive Methodology for Ecosystem Sustainability and Health (AMESH): An Introduction. In G. Midgley and A. Ochoa-Arias (eds.), *Community Operational Research: Systems Thinking for Community Development*. Boston: Kluwer Academic Publishers.

White, M. and Epston, D. (1990). *Narrative Means to Therapeutic Ends*. New York: WW Norton & Company.

WHO (1986). *Ottawa Charter for Health Promotion*. A declaration of the First International Conference on Health Promotion, Ottawa, Canada, 17–21 November 1986, published by the WHO Regional Office at http://www.who.dk./policy/ottawa.htm

Wobeser, G. A. (1994). *Investigation and Management of Disease in Wild Animals*. New York: Plenum Press, 265 pp.

Woo, D. and Vicente, K. (2003). Sociotechnical systems, risk management, and public health: comparing the North Battleford and Walkerton outbreaks. *Reliability and Engineering System Safety* 80: 253–269.

Index